Fei Ai

Slope stability along active and passive continental margins

Fei Ai

Slope stability along active and passive continental margins

Südwestdeutscher Verlag für Hochschulschriften

Impressum / Imprint
Bibliografische Information der Deutschen Nationalbibliothek: Die Deutsche Nationalbibliothek verzeichnet diese Publikation in der Deutschen Nationalbibliografie; detaillierte bibliografische Daten sind im Internet über http://dnb.d-nb.de abrufbar.
Alle in diesem Buch genannten Marken und Produktnamen unterliegen warenzeichen-, marken- oder patentrechtlichem Schutz bzw. sind Warenzeichen oder eingetragene Warenzeichen der jeweiligen Inhaber. Die Wiedergabe von Marken, Produktnamen, Gebrauchsnamen, Handelsnamen, Warenbezeichnungen u.s.w. in diesem Werk berechtigt auch ohne besondere Kennzeichnung nicht zu der Annahme, dass solche Namen im Sinne der Warenzeichen- und Markenschutzgesetzgebung als frei zu betrachten wären und daher von jedermann benutzt werden dürften.

Bibliographic information published by the Deutsche Nationalbibliothek: The Deutsche Nationalbibliothek lists this publication in the Deutsche Nationalbibliografie; detailed bibliographic data are available in the Internet at http://dnb.d-nb.de.
Any brand names and product names mentioned in this book are subject to trademark, brand or patent protection and are trademarks or registered trademarks of their respective holders. The use of brand names, product names, common names, trade names, product descriptions etc. even without a particular marking in this work is in no way to be construed to mean that such names may be regarded as unrestricted in respect of trademark and brand protection legislation and could thus be used by anyone.

Coverbild / Cover image: www.ingimage.com

Verlag / Publisher:
Südwestdeutscher Verlag für Hochschulschriften
ist ein Imprint der / is a trademark of
OmniScriptum GmbH & Co. KG
Heinrich-Böcking-Str. 6-8, 66121 Saarbrücken, Deutschland / Germany
Email: info@svh-verlag.de

Herstellung: siehe letzte Seite /
Printed at: see last page
ISBN: 978-3-8381-5063-5

Zugl. / Approved by: Bremen, UB, Diss., 2013

Copyright © 2015 OmniScriptum GmbH & Co. KG
Alle Rechte vorbehalten. / All rights reserved. Saarbrücken 2015

Table of Contents

Abstract 1

Zusammenfassung 3

1 Introduction 7
   1.1 Motivation 7
   1.2 Submarine slope stability processes 9
      1.2.1 Types of submarine mass movements 9
      1.2.2 Preconditioning factors and triggering mechanisms of submarine mass movements 11
      1.2.3 Geotechnical study of submarine mass movements 13
   1.3 Regional setting of case studies: passive vs. active continental margins 16
      1.3.1 Uruguayan and northern Argentine margin: passive margin 18
      1.3.2 Gela basin, central Mediterranean continental margin: foreland basin-active margin 20
      1.3.3 Ligurian basin, southeastern French continental margin: back-arc basin-active margin 22
   1.4 Methods 24
      1.4.1 Geophysical methods 24
      1.4.2 Coring and sedimentological methods 24
      1.4.3 Geotechnical methods 26
   1.5 Outline of the thesis 31
   References 32

2 Geotechnical characteristics and slope stability along the Uruguayan and northern Argentine margin 41
   2.1 Introduction 42
   2.2 Regional geological, morphological and oceanographic settings 42
   2.3 Material and methods 45
      2.3.1 Shipboard tests 46
      2.3.2 Laboratory tests 46
      2.3.3 Overpressure estimation 46
      2.3.4 Slope stability assessment 47
      2.3.5 Prediction of peak ground acceleration (PGA) 48
   2.4 Results 49
      2.4.1 Physical and geotechnical properties of sediments 49
      2.4.2 Slope stability analysis 54
   2.5 Discussion 56
      2.5.1 Preconditioning factors 56
      2.5.2 Triggering mechanisms 57
      2.5.3 Slope failure modes between open slope versus canyon 58
      2.5.4 Slope failures along the Uruguayan and northern Argentine margin versus slope failures on other passive margins 59
   2.6 Conclusions 60
   Acknowledgements 60
   References 60

3 Submarine slope stability assessment of the central Mediterranean continental margin: the Gela Basin 65

    3.1 Introduction 65

    3.2 Geological setting 66

    3.3 Material and methods 66

        3.3.1 Shipboard and laboratory analysis 66

        3.3.2 Overpressure estimation 68

        3.3.3 Slope stability analysis 68

    3.4 Results 69

        3.4.1 Physical and geotechnical properties 69

        3.4.2 Slope stability analysis 70

    3.5 Discussion 71

        3.5.1 Preconditioning factors 71

        3.5.2 Triggering factors 72

    3.6 Conclusions 73

    Acknowledgements 73

    References 73

4 Geotechnical characteristics and slope stability analysis on the deeper slope of the Ligurian margin, Southern France 77

    4.1 Introduction 78

    4.2 Regional geological, morphological and oceanographic settings 78

    4.3 Material and methods 80

        4.3.1 Laboratory tests 80

        4.3.2 Slope stability analysis 80

    4.4 Results 81

        4.4.1 Physical and geotechnical properties 81

        4.4.2 Slope stability analysis 84

    4.5 Discussion 85

        4.5.1 Preconditioning factors of WS and ES: superficial failure vs. deep-seated failure 85

        4.5.2 The influence of earthquake to the slope stability 86

    4.6 Conclusions 87

    Acknowledgements 87

    References 87

5 Conclusion and outlook 89

    5.1 Conclusion 89

    5.2 Outlook 90

    References 91

Acknowledgements 93

Appendix A: Core descriptions, physical and geotechnical properties of Uruguayan and northern Argentine margin 95

Appendix B: Core descriptions, physical and geotechnical properties of Gela Basin 113

Abstract

Submarine mass movements are widespread at submarine slopes and play an important role in transporting sediments across the continental slope to the deep basin, as well as potential danger to both offshore infrastructures (e.g., pipeline, cables and platforms) and coastal areas (e.g., slope failure-induced tsunamis). Sliding of the sediments on continental slope takes place when the shear stress within sediments exceeds the shear strength thereby causing slope failure. Slope failures are generally controlled by long-term preconditioning factors (e.g., high sedimentation rate, weak layer and oversteepening) and short-term triggering mechanisms (e.g., earthquake, anthropogenic activity). However, the exact causes for the different slope failure styles are still poorly understood.

In summary, this thesis investigates preconditioning factors and triggering mechanisms governing slope instabilities of three distinct submarine landslides areas in passive and active continental margin settings. Geotechnical properties of sediments from undeformed, headwall and deposits present different stress histories and shear strengths (undrained and drained shear strength). Geotechnical results are used for infinite slope stability of undeformed sediments under various conditions (undrained and drained, each static and earthquake conditions) to identify the preconditioning factors and quantify the influence of earthquakes as a key factor in slope failing mechanisms. The three distinct case studies are located at: (1) the passive continental slope of Uruguay and north of Argentina, (2) the low seismic and tectonically active Gela foreland basin, central Mediterranean continental margin, and (3) the moderate seismic and tectonically active back-arc basin, deeper slope of the Ligurian margin, Southern France.

1) On the Uruguayan and northern Argentine slope, submarine mass movements are common primarily because of high fluvial discharge by the Rio de la Plata River and strong bottom current forces within the Brazil-Malvinas Current Confluence (BMC) zone. Sedimentological, physical and geotechnical results of core samples were investigated to quantatively assess slope stability for two distinct study areas: mass movements dominated area off Uruguayan slope called Northern Slide area (NS) and canyons dominated area off the Rio de la Plata River mouth of Northern Argentine slope called Southern Canyon area (SC). NS mainly consists of clayey silt with interbedded sand layers with wide changes of physical and geotechnical properties from surficial (0-3 m) to deeper sediments (> 3 m): bulk density (1.5-2.1 g/cm^3), water content (20-95%), void ratio (0.6-3.0) and undrained shear strength (5-200 kPa from 0 to 16 m below seafloor (mbsf)). In contrast, SC mainly contains silty sand with high bulk density (1.7-2.4 g/cm^3), low water content (20-40%), low void ratio (0.6-1.2) and low undrained shear strength (5-20 kPa from 0 to 20 mbsf). Oedometer tests of both sites show overconsolidated (overconsolidation ratio, OCR: 1.5-12.7) near the seafloor and underconsolidated (OCR: 0.13-0.2) at depths of 20-30 mbsf and direct shear tests indicate that NS materials have a lower angle of internal friction (30.3-34.3°) compared to those of SC (36.9-41.3°). Slope stability analysis suggest that NS is sufficiently stable and is unlikely to experience repeated small-scale slope failures under the current conditions, but may experience unstable conditions if external triggers (e.g., earthquakes) are strong enough to trigger slope failure. In contrast, low stability of SC's steep slopes is reflected by repeated small-scale slope failures both during static conditions and certainly during seismic events.

2) On the low seismic and tectonic active continental slope of the Gela Basin in the Sicily Channel, central Mediterranean, high sedimentation rates and seismic loading seem to be the most important factors to initiate submarine landslides in Holocene. One relative small scale (5.7 km^2, 0.6 km^3), 8 kyr

old landslide named Northern Twin Slide (NTS) was studied through geotechnical measurements of sediments from undeformed upper slope recovered by MeBo corer (MARUM seafloor drill rig). The NTS region is characterized by two prominent failure scars that form two morphological steps of 110 m and 70 m height. Sediments show fine grain size (high clay content), high water content, low undrained shear strength and low internal friction angle, all of which suggests a weak layer around 28-45 mbsf that may act as potential slip surface in a future failure event. Oedometer tests attest the sediments are highly underconsolidated and the average overpressure ratio λ^* is ~0.7. Slope stability analyses indicate that the slope is stable under both static undrained and drained conditions. It suggests that moderate seismic triggers may have been responsible for the Northern Twin Slide formation and could also cause mass wasting in the future.

3) On the moderate seismic and tectonic active continental slope of Ligurian margin, northwestern Mediterranean Sea, submarine slope failures of various types and sizes are prevalent primarily because of seismicity up to ~M6, rapid sediment deposition in the Var fluvial system, and steepness of the continental slope (average 11°). Geophysical, sedimentological and geotechnical results of two distinct slides in water depth >1500 m: one located on the flank of the Upper Var Valley called Western Slide (WS), another located at the base of continental slope called Eastern Slide (ES) were studied to quantitatively assess slope stability. WS is a superficial slide characterized by a slope angle of ~4.6° and shallow scar (~30 m) whereas ES is a deep-seated slide with a lower slope angle (~3°) and deep scar (~100 m). Both areas mainly comprise clayey silt with intermediate plasticity, low water content (30-75 %) and under-consolidation to strong overconsolidation. Upslope undeformed sediments have low undrained shear strength (0-20 kPa) increasing gradually with depth, whereas an abrupt increase in strength up to 200 kPa occurs at a depth of ~3.6 m in the headwall of WS and ~1.0 m in the headwall of ES. These boundaries are interpreted as earlier failure planes that have been covered by hemipelagite or talus from upslope after landslide emplacement. Infinite slope stability analyses indicate both sites are stable under static conditions; however, slope failure may occur in undrained earthquake condition. Different failure styles include rapid sedimentation on steep canyon flanks with undercutting causing superficial slides in the west and an earthquake on the adjacent Marcel fault to trigger a deep-seated slide in the east.

Overall, the geotechnical investigations of these three case studies imply that (i) grain size of sediments seems to be the main factor affecting the physical and geotechnical properties, (ii) seismic loading is an important trigger mechanism even on passive and low seismic active continental margin, (iii) different slope failure types in adjacent areas are mainly controlled by grain size variation, sedimentation rate, rework of bottom currents and distance to the epicenter of earthquake. Geotechnical investigation combined with slope stability and earthquake analysis have well applied on quantitative estimation the preconditioning factors and triggering mechanisms of slope failure on both passive and active continental margins, and this method can be applied to other continental margins worldwide as well. An implementation of this geotechnical approach may require the basic geotechnical parameters of sediments such as bulk density, grain density, shear strengths under both drained and undrained conditions with the slope geometric parameters (slope angle and slope failure depth). Once the right parameters are chosen, the slope stability analysis can be used for evaluation of the stable states of slope under various scenarios at present and prediction potential slope failure modes in future.

Zusammenfassung

Submarine Massenbewegungen sind weit verbreitet entlang submariner Hänge. Sie spielen eine wichtige Rolle beim Sedimenttransport über den Kontinentalhang in die Tiefsee, stellen aber auch eine potentielle Gefahr für offshore Infrastrukturen (z.B. Pipelines, Kabel und Plattformen) und Küstenregionen (z.b. durch Hangrutschungen verursachte Tsunamis) dar. Ein Abrutschen von Sediment vom Kontinentalhang findet dann statt, wenn die Scherspannung im Sediment dessen Scherfestigkeit übersteigt und so ein Hangversagen verursacht. Hangversagen wird im Allgemeinen von Langzeit-Faktoren (z.B. hohe Sedimentationsraten, schwache Schichten, Übersteilung) und auslösenden Kurzzeit-Mechanismen, auch Trigger, (z.B. Erdbeben, anthropogene Aktivitäten) kontrolliert. Die genauen Ursachen für die unterschiedlichen Arten von Hangrutschungen sind jedoch bislang unzureichend verstanden.

Diese Doktorarbeit untersucht die Langzeit-Faktoren und Trigger von Hangrutschungen entlang passiver und aktiver Kontinentalränder in drei unterschiedlichen Erkundungsgebieten. Geotechnische Eigenschaften von oberhalb (z.b. ungestörten Sedimentproben) und unterhalb (z.b. gestörten Sedimentproben) der Abrisskante abgelagerten Sedimenten spiegeln verschiedene Spannungspfade (Belastungsgeschichten) und Scherfestigkeiten (dräniert und undräniert) wider. Die geotechnischen Ergebnisse fließen in infinite Hangstabilitätsanalysen von undeformierten Sedimenten unter verschiedenen Bedingungen (undräniert und dräniert, jeweils unter statischen und Erdbebenbedingungen) ein, um die bedingenden Langzeit-Faktoren zu identifizieren und den Einfluss von Erdbeben als Schlüsselfaktor von Mechanismen, die zu Hangversagen führen zu quantifizieren. Die drei unterschiedlichen Studien liegen entlang des: (1) passiven Kontinentalhangs von Uruguay und Nord-Argentinien, (2) schwach seismischen und tektonisch aktiven Gela-Beckens, zentral gelegen am Kontinentalrand im Mittelmeer und (3) moderat seismischen und tektonisch aktiven back-arc Beckens am tieferen Hang des Ligurischen Kontinentalrands, Süd-Frankreich.

1) Entlang des uruguayischen und nord-argentinischen Kontinentalhang kommen submarine Massenbewegungen aufgrund des hohen Abflusses des Rio de la Plata und der starken Kräfte der Bodenströmung in der Brazil-Malvinas Current Confluence (BMC) Zone häufig vor. Sedimentologische, physikalische und geotechnische Ergebnisse aus Kernproben wurden untersucht um die Hangstabilität zweier unterschiedlicher Studiengebiete zu quantifizieren: ein von Hangrutschungen dominiertes Gebiet auf dem uruguayischen Kontinentalhang, das Northern-Slide-Gebiet (NS), und ein von Canyons dominiertes Gebiet vor der Rio de la Plata Mündung auf dem nord-argentinischen Kontinentalhang, das Southern-Canyon-Gebiet (SC). Das NS besteht hauptsächlich aus tonigem Silt mit eingebetteten Sandlagen, dessen physikalische und geotechnischen Eigenschaften sich stark von oberflächlichen (0-3 m) zu tieferen Sedimenten (> 3 m) hin verändern. Folgende Wertebereiche für Teufen von 0-16 m unterhalb des Meeresbodens ließen sich für die unterschiedlichen Indexparameter ermitteln: 1) Dichte. 1.5-2.1 g/cm^3, 2) Wassergehalt: 20-95%, 3) Porenzahl: 0.6-3.0 und 4) undränierte Scherfestigkeit: 5-200 kPa. Im Gegensatz dazu besteht das SC hauptsächlich aus siltigem Sand mit höherer Dichte (1.7-2.4 g/cm^3), geringerem Wassergehalt (20-40%), geringerer Porenzahl (0.6-1.2) und geringerer undränierter Scherfestigkeit (5-20 kPa) für eine Teufebis 20 m. Ödometertests an Sedimenten aus beiden Gebieten zeigen eine Überkonsolidierung (Konsolidierungsgrad OCR 1.5-12.7) in Meeresbodennähe und eine Unterkonsolidierung (OCR 0.13-0.2) in Teufen von 20-30 m. Direktscherversuche ergeben, dass Material aus dem NS einen

geringeren inneren Reibungswinkel (30.3-34.3°) besitzt als Sedimente des SC (36.9-41.3°). Analysen der Hangstabilität ergeben, dass das NS stabil genug ist und wiederholte kleinskalige Hangrutschungen unter gegenwärtigen Verhältnissen unwahrscheinlich sind, aber instabile Verhältnisse eintreten können, sollten externe Trigger (z.B. Erdbeben) stark genug sein. Im Gegensatz dazu wird die geringe Stabilität der steilen Hänge des SCs durch wiederholte kleinskalige Hangrutschungen sowohl während statischer, auf jeden Fall aber während seismischer Ereignisse widergespiegelt.

2) Entlang des gering seismisch und tektonisch aktiven Kontinentalhangs des Gela-Beckens im Kanal von Sizilien im zentralen Mittelmeer, scheinen-hohe Sedimentationraten und seismische Belastung die Hauptauslöser für submarine Hangrutschungen im Holozän zu sein. Eine relativ kleinskalige (5.7 km², 0.6 km³), 8000 Jahre alte Hangrutschung, die Northern-Twin-Rutschung (NTS), wurde anhand geotechnischer Untersuchungen an Sediment des unverformten oberen Hangabschnitts, das mit dem Meeresboden-Bohrgerät MeBo gewonnen wurde, studiert. Die NTS-Region wird von zwei ausgeprägten Rutschungsnarben charakterisiert, die zwei morphologische Stufen von 110 m und 70 m bilden. Die Sedimente zeigen einen hohen Tongehalt, hohen Wassergehalt, geringe undränierte Scherfestigkeit und geringen inneren Reibungswinkel, was zusammengenommen auf eine schwache Schicht in 28-45 mbsf deutet, die bei einem zukünftiges Versagen als potentielle Gleitfläche fungieren kann. Ödometertests bestätigen, dass die Sedimente deutlich unterkonsolodiert (OCR: ~0.7) sind. Analysen der Hangstabilität deuten darauf hin, dass der Hang sowohl unter statischen undränierten als auch statisch dränierten Verhältnissen stabil ist. Moderat seismische Trigger könnten verantwortlich für die Northern-Twin-Rutschung gewesen sein und auch zukünftige Massenbewegungen verursachen.

3) Entlang des moderat seismisch und tektonisch aktiven Ligurischen Kontinentalhangs im nordwestliches Mittelmeer sind Hangrutschungen unterschiedlicher Art und Ausmaße aufgrund seismischer Aktivität bis ~M6, rascher Sedimentation im fluvialen System der Var und der großen Neigung des Kontinentalhangs (durchschnittlich 11°) weit verbreitet. Es wurden geophysikalische, sedimentologische und geotechnische Ergebnisse von zwei unterschiedlichen Rutschungen in Wassertiefen > 1500 m untersucht, um die Hangstabilität quantitativ zu erfassen. Die Westliche Rutschung (WS) liegt an der Flanke des oberen Var-Tals, die Östliche Rutschung (ES) liegt am Fuß des Kontinentalhangs. Die WS ist eine oberflächliche Rutschung, die durch eine Hangneigung von ~4.6° und einer flachen Narbe von ~30 m charakterisiert wird, während die ES eine tiefsitzende Rutschung mit einer flacheren Hangneigung von ~3° und einer tiefen Narbe von ~100 m ist. Beide Gebiete werden hauptsächlich aus tonigem Silt aufgebaut, der eine mittlere Plastizität und geringen Wassergehalt (30-75 %) besitzt und unter- bis stark überkonsolidiert ist. Hangaufwärts liegende, undeformierte Sedimente besitzen eine geringe undränierte Scherfestigkeit (0-20 kPa), die mit der Tiefe zunimmt. In einer Tiefe von ~3.6 m unterhalb der Abrisskante beim WS und ~1.0 m unterhalb der Abrisskante beim ES nimmt die Scherfestigkeit plötzlich auf über 200 kPa zu. Interpretiert werden diese Übergänge als frühere Gleitflächen, die nach der Rutschung durch hemipelagische Sedimente oder Hangschutt bedeckt wurden. Infinite Hangstabilitätsanalysen ergeben, dass beide Gebiete unter statischen Verhältnissen stabil sind; unter undränierten Verhältnissen könnte jedoch während Erdbeben ein Hangversagen eintreten. Unterschiedliche Arten des Versagens schließen im Westen rasche Sedimentation auf den steilen Flanken des Canyon ein, wo Unterspülungen zu oberflächlichen Rutschungen führen und im Osten Erdbeben entlang der angrenzenden Marcel-Verschiebung, die tiefsitzende Rutschungen auslösen.

Im Gesamten implizieren die geotechnischen Untersuchungen der drei Fallstudien, dass (i) die Korngröße der Sedimente der Hauptfaktor zu sein scheint, der die physikalischen und geotechnischen

Zussamenfassung

Eigenschaften bestimmt, (ii) Belastung durch Erdbeben sogar entlang passiver und gering seismisch aktiver Kontinentalränder wichtige Trigger darstellen und (iii) die unterschiedlichen Arten von Hangversagen in angrenzenden Gebieten hauptsächlich durch Korngrößenvarianz, Sedimentationsraten, Aufarbeitung durch Bodenströmung und Entfernung zum Epizentrum eines Erdbebens kontrolliert werden. Geotechnische Untersuchungen, zusammen mit Hangstabilitäts- und Erdbebenanalysen stellen eine geeignete Methode dar, um bedingende Faktoren und Triggermechanismen für Hangversagen sowohl an passiven als auch an aktiven Kontinentalrändern quantitativ abzuschätzen. Diese Methode kann für Kontinentalränder weltweit angewandt werden. Die Anwendung dieser geotechnischen Näherung benötigt geotechnische Basisparameter wie Dichte, Korndichte, Scherfestigkeiten unter dränierten und undränierten Bedingungen zusammen mit Parametern zur Hanggeometrie wie Neigungswinkel und Tiefe der Gleitfläche. Nach Wahl der richtigen Parameter, kann eine Analyse der Hangstabilität dafür verwendet werden, um stabile Zustände eines Hanges während unterschiedlicher gegenwärtiger Szenarien zu bewerten und potentielle Hangversagen für die Zukunft vorherzusagen.

(Translated by Franziska Hellmich, Alois Steiner and Andre Hüpers)

1 Introduction

1.1 Motivation

In recent years, increasing use of state-of-the-art technologies in geophysical exploration and in-situ measurements have significantly advanced our understanding of the phenomena of submarine mass movements and their consequences (Chiocci et al 2011; Masson et al. 2010; Sultan et al. 2004). However, the interplay between the various geological factors controlling submarine mass movements is still not fully understood (Locat and Lee 2002). Submarine slope failures are one of the main processes for long distance sediment transport from land to deep sea and for shaping seafloor morphology. In addition, submarine landslides are gained wide attention because of their catastrophic impacts on both coastal areas (e.g., slope failure-induced tsunamis) and also offshore infrastructures (e.g., pipeline, cables and platforms) (Locat and Lee 2002).

In marine environments, the major geological hazards include earthquakes, volcanic eruptions, submarine landslides and secondary effects such as tsunamis (either triggered by earthquakes or landslides) (Chiocci et al. 2011; Fig. 1.1). Given 30% of World's population lives with 60 km of the coast, the hazard posed by submarine landslide generated tsunamis may cause several consequences (Yamada et al. 2012). The role of submarine landslide in producing tsunamis has been more recognized over the last 30 years following the studies of the 1929 Grand Banks event (Piper et al. 1999), the 1998 M 7.0 Sissano earthquake that occurred off the northern coast of Papua New Guinea (Bardet et al. 2003) and the tsunami induced by the 8150-year BP Storegga Slide (Bondevik et al. 2005). Offshore industry activities such as the Ormen Lange study have improved the understanding of the submarine landslide generated tsunami (Solheim et al. 2005). The tsunami is most likely to have been generated by rotational slides where a thick slide block with steep headwall can move rapidly downwards (Masson et al. 2006).

As one of important marine geological hazard, submarine mass movements have severe impact on human lives and offshore infrastructures (Fig. 1.1). Just for the offshore hydrocarbon industry, the cost of damage to pipelines caused by submarine mass movements is ~$400 million each year (Mosher et al. 2010). One of most famous examples is the 1929 Grand Banks event on the continental slope south of the island of St Pierre. An M 7.2 earthquake located 250 km south of Newfoundland led to a series of small regressive slumps which resultant turbidity current broken twelve submarine transatlantic cables (Piper et al. 1999). Following the event, the scientific community began to discuss what caused the cable breaks and how the transformation of mass movements after earthquake loading (Heezen and Ewing 1952). Another famous example is the Nice Airport slide which also generated a debris flow and turbidity current downslope that broke several telecommunication cables (Mulder et al. 1997). The slide was caused by a sudden overloading from landfilling operation and static liquefaction of loose sand layers present below the harbor embankment (Seed et al. 1988). With more new submarine landslides are discovered after the 1929 grand Banks event, many new models of submarine slope failure initiation were developed (e.g., Morgenstern 1967; Lee and Edwards 1986; Hampton et al. 1996; Dugan and Flemings 2000; Sultan et al. 2004; Kvalstad et al. 2005; Stegmann et al. 2007; Stigall and Dugan 2010; Strasser et al. 2011).

1 Introduction

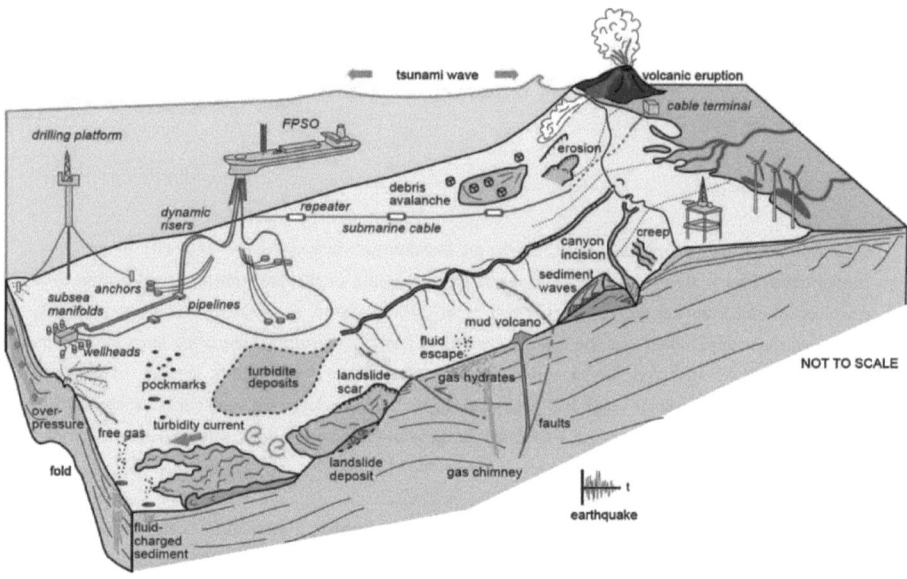

Fig 1.1 *Cartoon summarizing the continental margin features linked to geological hazard processes and main anthropogenic structures lying on the seafloor (taken from Chiocci et al. 2011).*

Understanding geological processes that govern and formation of submarine slope failures and subsequent potential hazards thus is a fundamental and societal relevant task of geologists and geotechnical engineers. To date, many questions about the complex mechanisms of slope stability on continental margins are still unresolved (Leynaud et al. 2009). It is well known that excess pore pressure within the sediment is a key parameter controlling consolidation and the shear strength of the deposited sediments, however, the evolution of excess pore pressure and how it relates to slope failure on continental margin is still no clearly understood. It is assumed that the internal structure of sediments plays an important role during slope stabilization. For instance, high porous ash layers have large potential of liquefaction and treat as potential slide planes (Harders et al. 2010), so that appropriate geotechnical measurements are needed to reliably determine the physical properties leading to failure. Many studies suggest seismic loading as a likely ultimate trigger mechanism, but quantitatively the effect of earthquake to slope instability is not well understood. To answer these questions, sediments of different types of mass movements were analyzed with laboratory geotechnical measurements to constrain the mechanics of mass movements. The geotechnical approach is proved to be a valuable tool to assess submarine mass movements in different tectonic areas (passive and active continental margins).

1.2 Submarine slope stability processes

1.2.1 Types of submarine mass movements

Sediments transport processes in marine environments are more complicated than those on land because of different driven forces including gravity, suspension, geostrophic circulation and so on (Fig. 1.2). Among them, submarine mass movements are the dominant processes transporting large amounts of sediment across continental slope to the deep ocean. Submarine mass movements are classified as slide (brittle deformation), slump (plastic deformation), debris flow (plastic or laminar flow) and turbidity current (fluid turbulent flow) according to the mechanical behavior of process (Mulder and Cochonat 1996; Mulder 2011; Fig. 1.2).

Submarine slides and slumps involve the movement of coherent masses of sediments bounded on all sides by distinct failure planes (Mulder et al. 1996). Differentiation of slides and slumps is based on the value of the Skempton ratio D/L (where D is the maximum depth of the slip surface, and L is the total length of the slump). Slides are translational with a D/L ratio generally < 0.15, whereas slumps are rotational and deep-seated with a D/L ratio between 0.15-0.33 (Skempton and Huchinson 1969). Most submarine slides appear to be translational and are characterized by a flat, slope-parallel basal failure surfaces (Canal et al. 2004). The failure surface is predetermined and corresponds to a discrete layer with low shear strength, such as permeable sand layers (e.g., Afen slide; Wilson et al. 2004), high porous tephra layers (e.g., submarine slide along the Middle American Trench; Harders et al. 2010) or sand and clay interbeds (e.g., hemipelagic and contouritic deposits offshore Norway; Laberg and Camerlenghi 2008). Submarine slides are commonly associated with a variety of extensional and compressional features. Extensional features are common in the upper part of the slide, especially in the headwall area, where listric faults are predominantly orientated perpendicular to the transport direction (Martinsen and Bakken 1990). In contrast, compression features are common at the front of the slide deposit where dominated by imbricate thrust slices of chaotically deformed strata (Martinsen 2005). Slides and slumps are not isolated processes and often form multiple phases of failures. Retrogressive slides are one of most common multiple phases of failure, which form because of upslope propagation of the failure (Mulder 2011).

Slides and slumps can transform to debris flows or turbidity currents through gradually increasing ambient water and disintegration of coherent blocks. Debris flows are flows in which the sediment is heterogeneous and may include lager clasts supported by a matrix of fine sediment (Lee et al. 2007). Turbidity currents involve the downslope transport of a relatively dilute suspension of sediment grains that are supported by an upward component of fluid turbulence (Parson et al. 2007). Turbidity currents are often formed by the disintegration of slides of debris flows, although they also may be generated independently of other gravity-driven processes (Mulder 2011).

Different types of submarine mass movements are represented different degrees of material transformation. To achieve it, sufficient energy was needed to reach a given degree of remolding (Fig. 1.3; Locat and Lee 2009). When the remolded energy available (E_{rA}) is larger than the remolded energy needed (E_{rN}), it is easily transferred to debris flows and turbidity currents. When smaller, it may tend to generate slides or slumps. However, the exact cause of transformation from slides to debris flows then to turbidity currents is still not fully understood. It is, however, out of the scope of this thesis, which focuses on the precondition factors and triggering mechanisms of initiation of submarine slope failure.

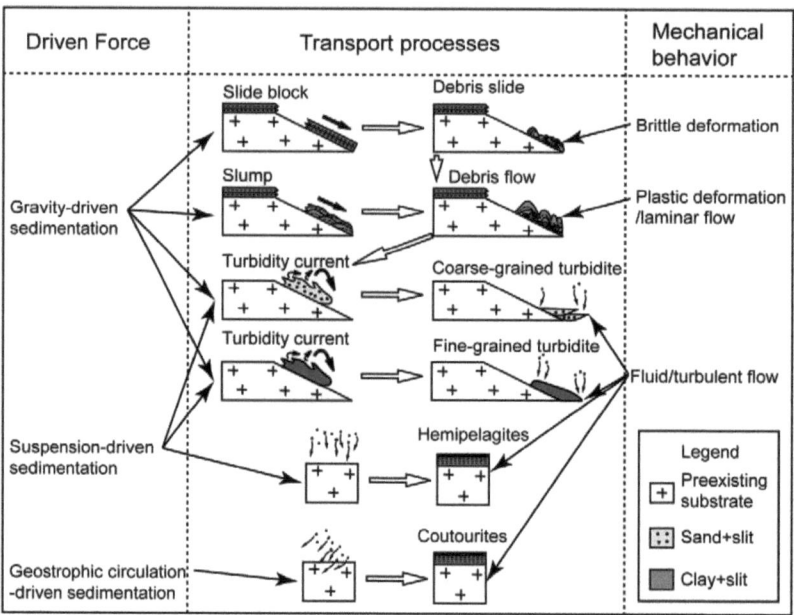

Fig. 1.2 Schematic diagram showing end-member types of gravity-driven, suspension-driven and geostrophic circulation-driven processes transporting sediments to the deep sea and their mechanical behaviors (Modified after McHugh et al. 2002; Madof et al. 2009)

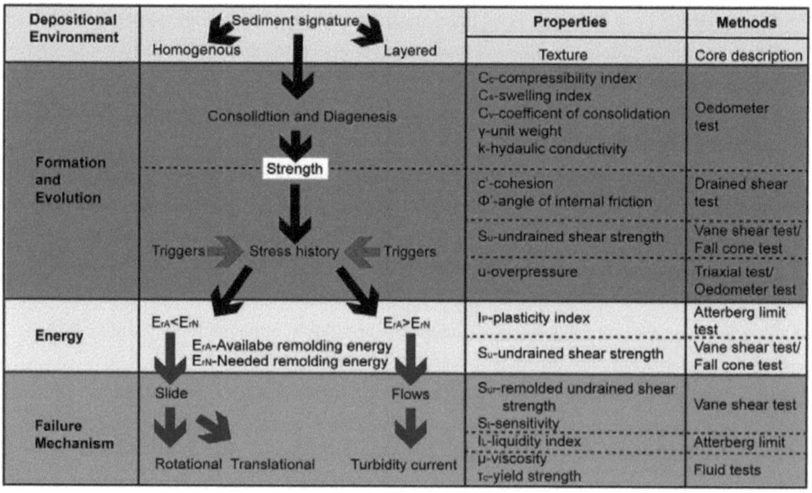

Fig. 1.3 Conceptual relationships between submarine sediment geotechnical properties, triggering mechanisms, and failure mechanisms (modified after Locat and Lee 2009; Locat and Lee 2002).

1.2.2 Preconditioning factors and triggering mechanisms of submarine mass movements

Sliding of the material downslope continental margins takes place when the shear stress within sediments exceeds the shear strength of the material thereby causing failure, which initiates the movement of materials downwards. Slope failure is generally controlled by long-term preconditioning factors and short-term triggering mechanisms (Leroueil 2001; Locat and Lee 2002; Sultan et al. 2004). Preconditioning factors are defined as the physical and geotechnical properties of sediments resulting from initial deposition and post-depositional alteration, which promote slopes susceptible to instability (Ercilla and Casas 2012). Preconditioning factors include the mass movement history, rapid sediment accumulation and under consolidation, the slope angle, the existence of a weak layer deposited over time and climate change over hundreds of years (Masson et al. 2006). The factors that cause submarine landslides, known as triggering mechanisms, are crucial to advancing the knowledge of processes of submarine landslides. Sultan et al. (2004) clearly defines a triggering mechanism as "an external stimulus that initiates the slope instability process." The triggering mechanisms include slope oversteepening, seismic loading, storm-wave loading, gas charging, gas hydrate dissociation, low tides, seepage, glacial loading and volcanic island processes (Locat and Lee 2002). Only a few submarine landslides exist for which exact trigger mechanisms are known with certainty (Mienert et al. 2003).

High sedimentation rates

Among various factors leading to a decrease in shear resistance of the sediments, the most important one is overpressure which can be generated by different mechanisms such as high sedimentation rates, gas charging, cyclic strength degradation, etc. Rapid sedimentation rates can generate overpressure in sediments and reduce the effective stress. Permeability, grain size, and structural arrangement of the sediment affect dissipation of overpressure in the sediments (Laberg and Camerlenghi 2008). In high latitudes, the climatic induced variability between glacial and interglacial situations seems to have been the main preconditioning factor for slope failure of northwestern European continental margin. Sedimentation rates are highest during glacial periods, with 36 and 65 m/kyr for Trænadjupet slide and Storegga slide, respectively (Laberg et al. 2003; Hjelstuen et al. 2004) whereas sedimentation rates are common less than 1 m/kyr during interglacials. Rapid loading of the deposits by unsorted glacial diamictons interlayered with high clay contents of 30-50% and low permeability of interglacial sediment resulted in build-up of overpressure and formation of an unstable sediment layer (Solheim et al. 2007). In low latitude, rivers discharge large quantities of sediment to relatively localized areas on the continental margins (Lee 2005). These thick, often underconsolidated sediments can fail even on very gentle slope angles (< 1°). High sedimentation rate is a primary mechanism of overpressure that facilitates large-scale slope failure in the Gulf of Mexico (Dugan and Sheahan 2012). Slope stability calculations for the Ursa region suggest that when sedimentation rates were 15 m/kyr, overpressures were generated but were still not driving the slope failures. When sedimentation rates were 30 m/kyr, high overpressures and flow focusing initiated failure in the upper 10 meter below sea floor (mbsf) (Dugan and Sheahan 2012).

Weak layers

Submarine landslides are often found to be rooted at one or more parallel-bedded sequences. These sequences likely represent a weak layer that plays an important role in landslide initiation. Weak layers as sedimentary units, which may fail under certain conditions include: sediment sequences able to maintain overpressure, high sensitivity, strain softening, clay-rich, high water content, and contactant behaviour (Solheim et al. 2007). In glaciated margins, weak layers have been identified in contouritic deposits that formed during interglacial periods and were rapidly buried under thick glacial marine deposits (Bryn et al. 2005). One or more slip planes of the Nyk slide offshore Norway are located within the contourite drift and parallel to the original acoustic lamination (Laberg et al. 2001). Geotechnical analysis suggests the clayey contouritic sediments have a high clay content, water content, plasticity index and liquidity index, resulting in a lower strength and higher sensitivity than the glacial sediments (Kvalstad et al. 2005). Compared to the clayey contouritic sediments, well sorted contouritic sands are inferred to have acted as detachment surfaces for the Afen slide on the continental slope in the Faroe-Shetland channel (Wilson et al. 2004). The Afen slide was initiated likely by liquefaction of the contouritic sands following seismic shaking (Wilson et al. 2004). Ash layers intercalated in the terrigenous or hemipelagic sediment sequences are also important for slope instability. During earthquake shaking, the grain framework of high permeable ash layer was collapsed which could produce overpressure and liquefaction of the layer. So, the ash layer has been suggested to serve as a detachment plane for translational sliding (Harders et al. 2010).

Earthquakes

Among the different possible driving forces, earthquake shaking is considered as the most important triggering mechanism for submarine landslide. Over 40% of submarine landslides were reported to be triggered by earthquakes (Hance 2003). Earthquakes can increase the driving stress by seismic accelerations and also may trigger sediment liquefaction of coarse-grained, cohesionless sediments (Sultan et al. 2004). Nadim (2012) presented three scenarios of earthquake-induced slope failure: (1) failure occurs during the earthquake, where the excess pore pressure generated by cyclic stresses that degrade the shear strength so much that the slope is not able to carry the static shear stress, (2) failure occurs post of earthquake, the excess pore pressures generated by the cyclic stresses migrate from deeper layer into critical areas leading to slope instability, (3) failure due to creep after earthquake, it is believed to be the most common mechanism for clay slopes. Earthquake is not only common to trigger submarine landslides in active continental margins, but also may play a critical role in initiating slope failure even in passive continental margins. The well-known Grand Banks submarine landslide was triggered by a magnitude 7.2 earthquake south of Newfoundland in 1929 (Fine et al. 2005). Postglacial earthquake activity due to glacio-isostatic rebound was considered the most likely triggering mechanism for Storegga slide offshore Norway (Kvalstad et al 2005). A simple method to account for earthquake on slope stability analysis will be presented following paragraphs assuming pseudostatic earthquake acceleration.

1.2.3 Geotechnical study of submarine mass movements

Geotechnical investigation of submarine mass movements is a direct approach to characterize physical and geotechnical properties of sediments for better understand causes and dynamics of slope failures. Sliding takes place when driving stress within sediments exceeds the resisting strength of the sediments (Fig. 1.4; Hampton et al. 1996). To understand the mechanics involved in slope failure, it is necessary to consider resisting strength and driving stress.

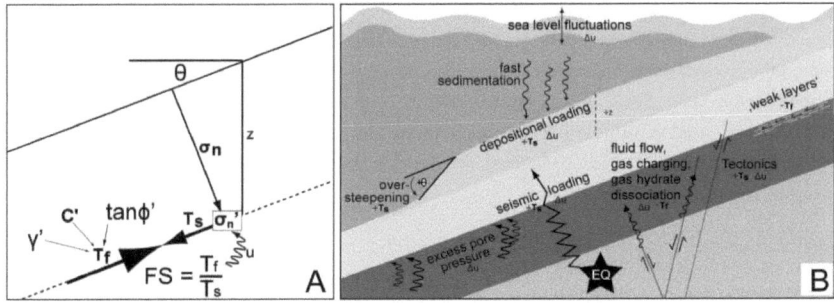

Fig. 1.4 *(A) Schematic diagram showing infinite slope stability analysis. (B) Schematic diagram showing how different geological processes affect the slope instability (taken from Strozyk 2009).*

Resisting and driving stress

The shear strength of sediment represents its ability to resist shear stress. As the shear stress applied to a sediment element steadily increases, the shear strain of the sediment element increase as well. At some point a limiting shear stress is reached which is taken as shear strength. If stresses are applied so rapidly that water cannot drain out, conditions are considered as undrained. In contrast, if stresses are applied so slowly that no excess pore pressures are developed, conditions are considered as drained. For drained condition, shear strength is expresses as a linear envelope obtained from the shear strength versus applied normal stress (Craig 2004):

$$\tau_f = c' + (\sigma_n - \Delta u)\tan\phi' \qquad (1.1)$$

Where τ_f is the shear strength, c' is the effective cohesion, σ_n is gravitational stress acting on the failure surface, Δu is overpressure, ϕ' is the effective angle of friction. Assuming slope angle is θ, σ_n is expressed as:

$$\sigma_n = \sigma'_{vh}\cos^2\theta = \gamma'z\cos^2\theta \qquad (1.2)$$

Where σ'_{vh} is vertical effective overburden stress in hydrostatic condition ($\sigma'_{vh} = \gamma'z$), γ' is buoyant weigh, z is overburden depth. For undrained conditions, the undrained shear strength is represented by:

$$S_u = \sigma'_v S \quad \text{(for normally consolidated sediment)} \qquad (1.3)$$

$$S_u = \sigma'_v S(OCR)^m \quad \text{(for overconsolidated sediment)} \qquad (1.4)$$

S_u is the undrained shear strength, σ'_v is the vertical effective overburden stress, S is a sediment constant (often equal to ~0.3 for fine-grained marine sediments, Lee and Edwards 1986), OCR is overconsolidation ratio, and m is a sediment constant commonly equal to ~0.8. The vertical effective stress is expressed following Terzaghi et al. (1996):

$$\sigma'_v = \sigma_v - u = \sigma - (u_0 + \Delta u) = (\rho_b - \rho_w)gz - \Delta u = \gamma'z - \Delta u \tag{1.5}$$

Where σ_v is total overburden stress, ρ_b is bulk density, ρ_w is water density, γ is unit weight of the bulk sample, γ_w is unit weight of water, and g is gravity acceleration.

From above equations of shear strength in drained and undrained conditions, it is found that overpressure is an important parameter to reduce the shear strength of the sediment. Overpressure can be generated in various ways, such as rapid sedimentation rates, gas hydrate dissociation, and cyclic stresses by earthquakes or storm waves. Overpressure is common to be generated by rapid sedimentation rates. Since insitu pore pressure of deep-sea sediment is difficult to measure, two methods were used to estimate overpressure generated by rapid sedimentation rates. Preconsolidation stress (σ'_{pc}) interpreted from oedometer tests is a simple approach for first-order estimation of overpressure (Casagrande 1936):

$$\Delta u = \sigma'_{vh} - \sigma'_{pc} \tag{1.6}$$

Overpressure due to sedimentation also can be evaluated with Gibson's (1958) one-dimensional solution under the assumption that a constant sedimentation rate and no flow at underlying strata occurs. The modeled overpressure is controlled by Gibson's time factor (T_g) (Flemings et al. 2008):

$$T_g = m^2 t / c_v \tag{1.7}$$

Where m is sedimentation rate, t is time, and c_v is coefficient of consolidation ($c_v = k/(m_v\gamma_w)$), the latter of which depends on coefficient of permeability (k) and coefficient of volume compressibility (m_v), both being obtained from oedometer tests.

In an infinite surface, the downslope gravitational shear stress (τ_s) at any point below the seafloor is given by:

$$\tau_s = \gamma'z\sin\theta \tag{1.8}$$

Infinite slope stability analysis

According to the low average slope angles (~2°) of submarine slopes and the low ratio between failure depth and spatial extent for submarine landslide, the infinite slope stability is assumed to be appropriate using to calculate the factor of safety (FS). Within this model, the seafloor is represented by a long, wide ramp with a uniform slope gradient (θ) and failure is assumed to occur along planes parallel to the slope surface. The factor of safety determines whether a given slope is stable (FS > 1) or susceptible to failure (FS ≤ 1). The shear strength of sediments depends on the conditions and time of drainage during shear. It is essential to consider long-term factors such as overpressure induced by sedimentation (drained condition) and short-term factors (undrained condition) such as forces induced by earthquakes. Slope stability was evaluated for four different scenarios:

(1) Static undrained conditions can be affected by rapid change in slope geometry or fluctuation of pore pressure. The factor of safety calculation after Morgenstern (1967) and Løseth (1999) follows:

$$FS = \frac{S_u}{\gamma'z\sin\theta\cos\theta} \quad \text{(Undrained static)} \quad (1.9)$$

(2) <u>Static drained conditions</u> respond to long-term steady state pore pressure (Dugan and Flemings 2002):

$$FS = \frac{c'+\gamma'z\left(\cos^2\theta - \lambda^*\right)\tan\phi'}{\gamma'z\sin\theta\cos\theta} \quad \text{(Drained static)} \quad (1.10)$$

Where λ^* is overpressure ratio ($\lambda^* = \Delta u/\sigma'_{vh}$).

(3) <u>Earthquake undrained conditions</u> use pseudostatic analysis for a simplified evaluation of the seismic factor of safety of a slope. In order to simplify the slope stability analysis, it is common to assume that only horizontal ground motion contributes to the slope failure. The earthquake force is represented by a horizontal force and a pseudostatic seismic coefficient (k_e). The pseudostatic acceleration (a) is k_e times the gravitational acceleration g (a = k_e g), which is assumed to be applied over a time period long enough for the induced shear stress to be considered being constant (Hampton et al 1996). The undrained pseudostatic factor of safety is given by the following expression (ten Brink et al. 2009):

$$FS = \frac{S_u}{\gamma'z\left[\sin\theta\cos\theta + k_e\left(\gamma/\gamma'\right)\cos^2\theta\right]} \quad \text{(Undrained earthquake)} \quad (1.11)$$

(4) <u>Earthquake drained conditions</u> only include pre-earthquake pore pressure (not considering the overpressure developed during seismic shaking) under the assumption that shear strength does not decrease during seismic shaking (Mulder et al. 1994):

$$FS = \frac{c'+\gamma'\left(\cos^2\theta - \lambda^*\right)\tan\phi}{\gamma'z\left[\sin\theta\cos\theta + k_e\left(\gamma/\gamma'\right)\cos^2\theta\right]} \quad \text{(Drained earthquake)} \quad (1.12)$$

Prediction of Peak ground acceleration

The critical pseudostatic acceleration (a_c) is the earthquake acceleration at which earthquake induced stress just equals the shear strength (FS = 1 of Equations 1.11 and 1.12). Critical pseudostatic acceleration as the average equivalent uniform shear stress imposed by seismic shaking represents ~65% of the effective seismic peak ground acceleration (PGA = a_c/65%) (Seed and Idriss 1971; Seed 1979; Strasser et al. 2011). The median ground motion of peak ground acceleration was estimated using empirical seismic attenuation relationships according to different areas (Bindi et al. 2011; Campbell and Bozorgnia 2008). The absolute value of PGA depends on magnitude, source distance, style of faulting of the earthquake, hanging-wall, site response and basin response.

1.3 Regional setting of case studies: passive vs. active continental margins

Submarine landslides occur frequently on both passive and active continental margins, especially on the continental slopes, where earthquakes and rapid deposition of sediments are the major causes of landsliding (Nelson et al. 2011; Locat and Lee. 2009). Figure 1.5 shows the distribution of the main known submarine mass movements of passive and active continental margins around world oceans.
Passive continental margins refer to tectonically divergent or tailing margins which are generally characterized by a smooth relief with wide continental shelves, low continental slope angles, moderate to high sedimentation rates and just local tectonic activities. The accumulation of thick sediments on flat continental slope is potentially hosting large volume and long run-out distance landslides. Offshore Norway, rapid loading by glacial sediments and weak layers related to interglacial contouritic drifts lead to many large submarine landslides, such as Storegga slide and Trænadjupet slide (Bryn et al. 2005; Laberg et al. 2002). In Gulf of Mexico, the rapidly deposited and organic-rich muddy deltaic sediments from the Mississippi River results in oversteepened, gas-charged and unstable slopes (Coleman 1988; Sawyer et al. 2009). Continental margin off northwestern Africa is characterized by oceanic upwelling as well as locally focused aeolian input resulting in relatively high sedimentation rates which leads to sediment instabilities (Krastel et al. 2006). In contrast, the continental margin off Uruguay and northern Argentina is dominated by strong contour currents in different depths and a high amount of fluvial sediment resulting in widespread contouritic deposits which potentially lead to small-scale and but more frequent landslides (Krastel et al. accepted; Krastel et al. 2011). In the western Mediterranean Sea, Messinian sea level lowstand caused deep erosion of many submarine canyons and resulted in slope instabilities on the flanks of canyon (Nelson et al. 2011).
Active continental margins refer to tectonically active convergent and transform-fault margins which commonly have narrow continental shelves, steep continental slopes and deep trenches. Submarine landslides on active margins generally have short run-out distances and small volumes of sediments. It is mainly because sediments become dense with time by frequent shaking of great earthquakes in active margins which resulting in seismic strengthening (Lee et al. 2004). Relatively dense sediments due to seismic strengthening of active margin also seldom mobilize into debris flows or turbidity currents compared to under-consolidated sediments of passive margins (Lee et al. 1991). Great earthquakes (> M8) along Cascadia and northern California cause seismic strength of the sediment resulting in minor mass transport deposits compared but not turbidites-system deposits (Nelson et al. 2011).
Here, geological settings of three case studies in this thesis are presented in detail: (1) Uruguayan and northern Argentine margin: passive margin, (2) Gela basin, central Mediterranean continental margin: foreland basin-active margin, (3) Ligurian basin, Southeastern French continental margin: back-arc basin-active margin.

1 Introduction

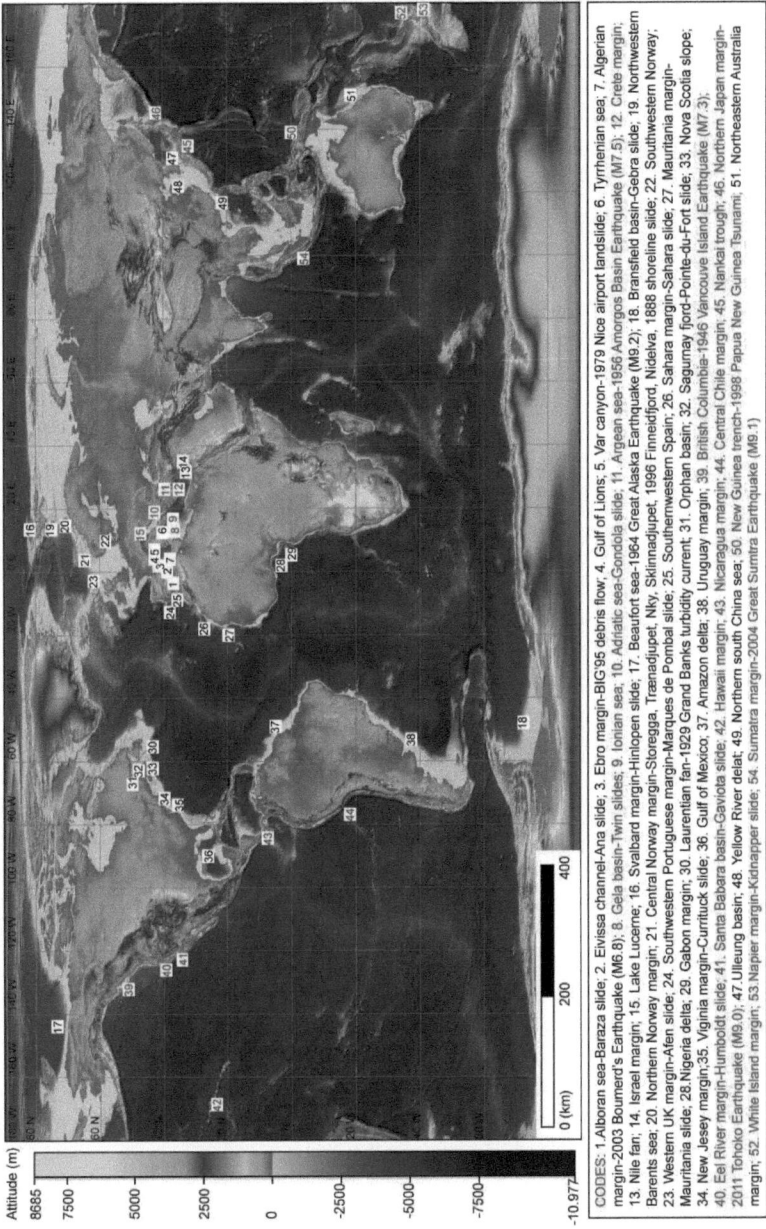

Fig. 1.5 Location map of worldwide distribution of submarine mass movements. Black numbers indicate mass movements occurred in passive margin. Red numbers indicate mass movements occurred in active margin.

1.3.1 Uruguayan and northern Argentine margin: passive margin

The Uruguayan and northern Argentine margin is considered as an extensive volcanic passive margin which formed during the opening of the South Atlantic in the early Cretaceous and is underlain by seaward dipping reflector (Hinz et al. 1999; Franke et al. 2007). Post-Cretaceous sedimentary sequences display major units separated by conspicuous seismic horizons (Hinz et al. 1999; Violante et al. 2010). Early Cenozoic was characterized by a high vertical accretion of the slope. Antarctic water masses began to play a significant role in forming prograding-retrograding sedimentary sequences and forming of submarine canyons in Eocene-Mid Miocene. Mid-Late Miocene was dominated by progradation and formation of the Ewing Terrace. Intensive contouritic and turbiditic activity final excavated the Mar del Plata Submarine Canyon in Late-Pliocene-Quaternary. The Argentine margin can be divided in four tectonic segments separated by transfer fracture zones (Franke et al. 2007). Our study area is located in the northernmost segment separated by the Salado transfer zone. Current tectonic activity in the study area is mainly characterized by active subsidence, which resulted in some intraplate seismic activities aligned along the Salado transfer zone (Sosa 1998). Recent documented earthquakes occurred in the years of 1849, 1888, and 1988 A.D. (for locations of epicenters, see Fig. 1.6, Sosa 1998). The epicenter and magnitude of the 1988 earthquake is not conclusively defined. Seismological observatory of the University of Brazil showed the epicenter located 36.5° S, 53.5° W, +/- 100 km with the regional magnitude of 3.9, whereas, NEIC (National Earthquake Information Center) recorded the epicenter of 36.27° S, 52.73° W with a body wave magnitude of 5.1 mb (Sosa 1998). Here we choose the seismic parameters as the epicenter of 36.27° S, 52.73° W with a body wave magnitude of 5.2 mb following Assumpção (1998).

The Uruguayan and northern Argentine margin is located in a key region of the world ocean that surface and deep Antarctic-sourced water masses interact with North Atlantic-sourced water masses (Piola et al. 2001). The oceanographic regime in the upper water column is characterized by the confluence of the Brazil and Malvinas Currents. The Malvinas Current (MC), a narrow branch of the Antarctic Circumpolar Current, flows northward along the Argentine margin up to approximately 37°S with volume transport range between 40-70Sv ($1Sv=10^6$ m^3/s) (Fig. 1.6; Matano et al. 2010). The Brazil Current (BC) is the western limb of the subtropical gyre that transports warm and salty waters towards the pole with a volume transport range between 25 to 40Sv at water depth to 500m (Matano et al., 2010). Both currents collide near the Rio de la Plata River estuary at 37°S; the merger of which creates the Brazil Malvinas Confluence (BMC). BMC turns southeast to transport warm water poleward and strongly governs sedimentary processes and morphology of the upper slope (Matano et al. 2010). Intermediate circulation below the confluence zone at 500-4000m water depth includes northward-flowing Antarctic Intermediate water (AAIW) and Upper and Lower Circum Polar Deep Water (CDW), as well as southward flowing North Atlantic Deep Water (NADW). At water depths below 4000 m, the deep circulation is dominated by Antarctic Bottom Water (AABW), which generates a strong cyclonic gyre trapped in the Argentine Basin (Flood and Shor 1988).

From Pliocene to Holocene, sedimentation along the Uruguayan and Argentine margin is strongly influenced by interaction of downslope and alongslope transport processes as indicated by the evolution of canyons, slope plastered drifts and channels (Krastel et al. 2011; Bozzano et al. 2011; Preu et al. 2012; Preu et al. 2013). The slope of Uruguayan margin is characterized by a smooth topography and gentle slopes (1-3°) typical of margins where deposition prevails over erosion (Ewing and Lonardi 1971). Several scarps and mass transport deposits (MTDs) at 1200-2800 m were found in NS (Krastel et al. 2011). The slope of the Northern Argentine margin is characterized by the flat Ewing terrace, which is dissected by canyons at ~1400 m (e.g., Mar del Plata Canyon and Querendi Canyon). The slope below the Ewing terrace is steep (3-7°), suggesting that erosion has been the main process shaping the slope (Ewing and Lonardi 1971). No clear MTDs were found in the sedimentary sequences upper slope and flanks of the Querendi Canyon. However, MTDs were imaged in the thalweg of the Querendi Canyon, which suggests headward erosion is common in this canyon (Krastel et al. 2011). The sediments of the study area are generally divided into a fine-grained dominated Uruguayan continental slope and coarse-grained dominated northern Argentine slope divided by the Brazil Malvinas confluence (BMC, ~37° S). North of the BMC, high quantities of terrigenous sediments (~80×10^6 ton/yr) discharged by Rio de la Plata River, containing 75% coarse to medium silt, 15% fine to very fine silt and 10% clay (Giberto et al. 2004), are swept northwards by alongshore currents and deposited at the Uruguayan continental shelf and slope (Piola et al. 2005). South of the BMC, coarse fluvial sediments are trapped with the estuary and occasionally carried directly down slope by turbidity current (Garming et al. 2005). On the deeper slope, Sediment transport on Uruguayan continental slope in water depth between 2000-4000 m is dominated by southward flow of North Atlantic Deep Water (NADW) whereas most of the surficial sediments at the northern Argentine slope have been carried and reworked northward by water masses in different depths (more details, see Piola and Matano 2001).

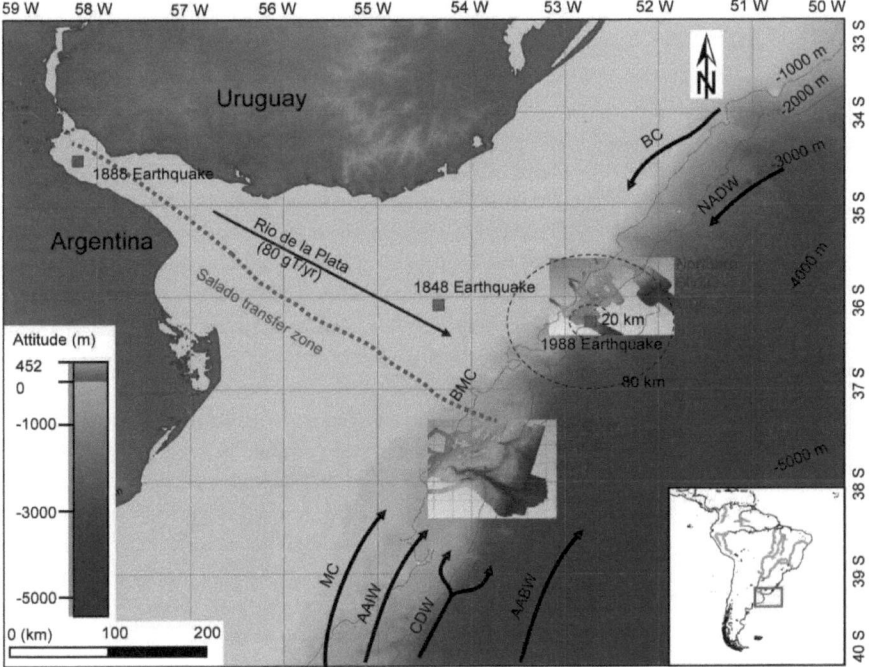

Fig. 1.6 *Map showing the location of study area and the oceanographic setting along the Uruguayan and northern Argentine margin. Red dots indicate the epicenters of the 1848, 1888 and 1988 earthquakes. MC: Malvinas Current, AAIW: Antarctic Intermediate Water, CDW: Circum Polar Deep Water, AABW: Antarctic Bottom Water, BC: Brazil Confluence, and NADW: North Atlantic Deep Water.*

1.3.2 Gela basin, central Mediterranean continental margin: foreland basin-active margin

The Gela Basin represents a Plio-Quaternary foredeep located at the front of the Maghrebian fold and thrust belt and is filled with 2.5 km of shallowing-upward marine sediments (Argnani 1990). In the north, the evolution of Gela Basin was connected to the emplacement of the Gela nappe, the southwest migrating outermost thrust wedge of the Maghrebian chain (Bulter et al. 1992 and Fig. 1.7). A general uplift has characterized the late Quaternary evolution of the thrust front area after the emplacement of the Gela nappe and volcanic activity has been widespread inside and outside the foredeep basin (Keller et al. 1978). Towards the south, the Gela foredeep has been affected by Plio-Quaternary extensional faulting related to an episode of crustal thinning which originated the NW-SE Trending Sicily channel rift zone (Reuther 1987). Compared with other Mediterranean area, Gela Basin is characterized in relatively low seismicity from historical records (USGS). Since 1970, at least seven epicenters of earthquakes, magnitude ranging 2.8 to 4, were registered in the study area (Fig. 1.7).

The marine circulation in the Sicily Channel is mainly driven by the water exchange between the Eastern and Western Mediterranean Sea. Sicily Channel is characterized by a two-layer system which maintained by excess evaporation in the Eastern Mediterranean. In the upper layer, the modified Atlantic water (MAW) flows from west to east whereas in the intermediate and deep layer, the Levantine Intermediate water (LIW) flows from east to west. Controlled by topographic features, the circulation in the channel forms meanders and eddies in variable strength, size and shape (Martorelli et al. 2011; Verdicchio and Trincardi 2008). The LIW forms a pair of subsurface eddies in Gela Basin, one cyclonic and the other anticyclonic, which reach velocities greater than 13 m/s (Lermusiaux and Robinson 2001 and Fig. 1.7).

The Messinian unconformity defines the base of the Plio-Quaternary sedimentary succession and is overlain by a ~250 m thick late-to-middle Pliocene sedimentary succession generally draping the basin floor (Minisini and Trincardi 2009). The uplift of the mainland region northeast of the Gela Basin generated strong erosion and nourished the westward dipping Quaternary progradational wedge during the mid-Pliocene to Quaternary (Gardiner et al. 1995). Sequence stratigraphic interpretation on the shelf and upper slope area of the Gela Basin (Minisini and Trincardi 2009) identify, from bottom to top: (1) A deep erosional unconformity (erosional surface ES2) truncating the seismic reflectors on the upper slope and becoming conformable in the deeper basin. The erosional surface ES2 has been interpreted as the shelf-wide erosional unconformity associated with the sea level lowstand during Marine Isotope Stage 6 (MIS6), (2) A sequence boundary (SB1), corresponding to MIS2 sea level fall, cutting through MIS3 deposits on the shelf, (3) A progradational wedge on the upper slope (Lowstand System Track, LST) interpreted as been formed during the last falling sea level leading to the glacial low stand, (4) A pronounced erosional surface (ES1) testifying subaerial exposure of the shelf and subsequent onset of the postglacial sea level rise in the basin. ES1 cuts the glacial progradational wedge and merge with SB1 on the shelf, while become conformable toward the basin, (5) A transgressive wedge (Transgressive System Track, TST), deposited during the drowning of the continental shelf, (6) A progradational wedge (Highstand System Track, HST) that started accumulating once the present sea level was reached, at around 5.5 kyr BP, (7) On the upper slope the youngest post-glacial sequence further is characterized by sediment drifts and associated slope-parallel moats, located along the 200 m contour (shelf-edge muddy contourite deposits) (Verdicchio and Trincardi 2008). The growth of contourite deposits on the slope of Gela Basin is consistent with the occurrence of a subsurface cyclonic gyre that flows along slope (Lermusiaux and Robinson 2001). Gela basin is characterized by widespread occurrence of repeated submarine mass movements. The gradually steeper of the progradational sequences of the Plio-Quaternary succession showed evidence of increasing slope instability as progradation proceeds (Minisini and Trincardi 2009). Gela Slide is the largest mass transport complex occurred at the base of the progradational units which has affected the northern margin of the basin mobilizing a volume of sediment in the order of 1000 km^3 in mid-Pleistocene times (Trincardi and Argnani 1990). Following this major basin-wide event, several small-scales slope failure events are exposed by fresh headwalls and a set of failed sediments up to the Holocene (Minisini et al. 2007) with recurrence intervals in the order of 3-4 ky since the Last Glacial Maximum (Minisini and Trincardi 2009). The main preconditioning factors of high recurrence of slope failure in the Gela Basin include: (1) an inherited westward dipping Mesozoic ramp, (2) increasing slope angles by rapid deposition of Quaternary units, (3) slope oversteepening induced by Father Slide, (4) rapid deposition of water-saturated and underconsolidated sediments by contour currents (Minisini et al. 2007; Verdicchio and Trincardi 2008).

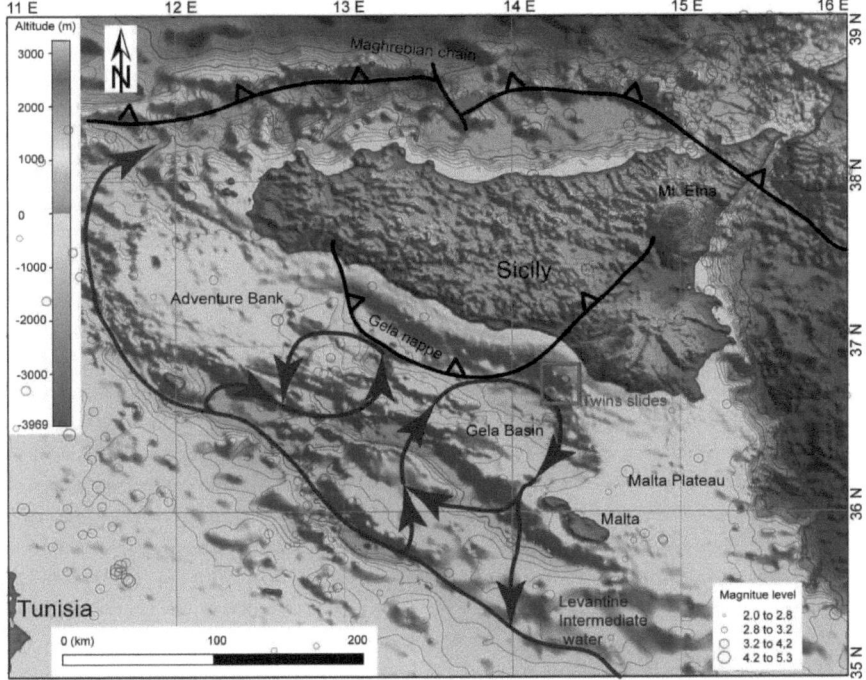

Fig. 1.7 *Map showing the location of the Gela Basin in the Sicily Channel, Central Mediterranean Sea. Blue lines indicate the bottom current in the central Mediterranean Sea (revised after Verdicchio and Trincardi 2008). Black lines indicate the position of frontal thrusts of Gela nappe and Maghrebides chain (taken from Catalano et al. 1996). Red circles indicate magnitude level of earthquakes recorded since 1970 (USGS database).*

1.3.3 Ligurian basin, Southeastern French continental margin: back-arc basin-active margin

Over the past 100 Myr, Ligurian Basin opened with the convergence between Africa and Eurasia plates (Dewey et al. 1989). Between 34 and 28 Myr, the continental started rifting and ended at around 21 Myr. Between 21 and 16 Myr, the continental Corsica-Sardinia block started counterclockwise rotation (Gattacceca et al. 2007). The Ligurian Basin is considered as a back-arc basin that formed by continental rifting and drifting during the late Oligocene from the southeastward rollback of the Apennines-Maghrebides subduction zone. This complex past geological history produced a particular complex topographic and structural domain on the southwestern Alps-Ligurian Basin junction at present (Larroque et al. 2012). The Alps-Ligurian Basin junction is one of the most seismically active areas in Western Europe. Currently, active basin deformation occurs offshore at a slow rate of ~1.1 m/ka NNW-SSE (Bethoux et al. 1998), which involves moderate seismic activity with common earthquake magnitudes of M2.2 to M4.5 (Rehault and Bethoux 1984; Fig. 1.8A). The focal depths of the earthquake are generally shallow with depth ranging 5 to 20 km (Courboulex et al. 1998). However, earthquake magnitudes up to M6.8 (e.g., 1887 Ligurian earthquake M_w 6.0; 1963 earthquake M_w 6.0) are documented in at the Ligurian margin (Larroque et al. 2012). The Marcel Fault shows evidence of present activity that three moderate earthquakes (M3.8-M4.6) took place around this fault over the last 30 years (Larroque et al. 2012 and Fig. 1.8).

The northern Ligurian margin is also one of the tsunamigenic zones of the northern Mediterranean Sea (Tinti et al. 2004). Two of the tsunamis that have occurred there are suspected to have been earthquake inducted: the 1887 event, and potentially the 1564 event (Larroque et al. 2012). Some other tsunamis have been generated by submarine slope failures, such as the 1979 Nice event (Labbé et al. 2012; Sahal et al. 2011).

Along the northern margin of the Ligurian Basin, the continental shelf is very narrow (2-3 km) or even absent offshore of Nice and in the Baie des Anges. The continental slope is steep, with a slope gradient of about 11°. The slope gradient decreases to ~3° at the boundary between continental slope and oceanic basin. The upper continental slope is eroded by two major canyons (Var canyon and Paillon canyon), which coalesce at a depth of 1650 m (Cochonat et al. 1993 and Fig. 1.8). A single channel was formed at the confluence of the two canyons and divided into three parts: upper, middle and lower valleys (Savoye et al. 1993). The During the Messinian crises, the Var paleocanyon was shaped and filled on the slope by a sea-level lowering of ~1500 m in the early Pliocene (~5.96-5.32 Ma) and the base of the slope was deposited of transgressive conglomerates and marls (Savoye et al. 1993). In the deep basin, it was responsible for the deposition of thick evaporate layers (Hsü et al. 1973). In the early Pliocene, hemipelagic clay was accumulated while coarse-grained sediment was trapped in the river Var area due to sea level high stand. In the middle Pliocene, Var canyon prograded in a steep Gilbert-type delta to the slope break and corresponded to the modern coastline (Savoye and Piper 1991). During the Quaternary, coarse-grained material was supplied to the slope by the braided river Var related to the coastal uplift and sediment transport from the glaciated Alps (Savoye et al. 1993).

At present, the Var and Paillon canyons are both link to the fluvial systems, which represent the most important erosive features (Klaucke et al. 2000) The Var river carries approximately 1.5×10^6 tons of particles per year that might generate high sedimentation rate (6.3-54 cm/yr) on the upper slope around the river mouth (Mulder et al. 1996). As a result, the sedimentary deposits on the upper slope are underconsolidated and likely to slide (Cochonat et al. 1993). Three major slope failures are observed:

(1) superficial slope failure, which occurs in area of low slope angles on the upper slope resulting from high sedimentation rate (Migeon et al. 2011), (2) canyon wall gullying by undercutting currents and debris flows at higher slope gradients resulting from oversteepening (Klaucke and Cochonat 1999; Klauke et al. 2000), and (3) deep-seated failures characterized by pronounced headwalls may resulting from seismic loading (Migeon et al. 2011). Large-scale slope failures are located at the base of continental slope. An impressive scar called Cirque Marcel is located at the base of the slope ,between 1300 and 2000 m of water depth and affect slope deposits over 100-300 m (Migeon et al. 2011; Fig. 1.8).

My study focuses on the deeper slope of Ligurian margin (Fig. 1.8). The walls of Upper Valley are highly dissected by small retrogressive failure events (Migeon et al., 2011) such as that west of Cap Ferrat Ridge called Western Slide (WS). It is characterized by shallow headwalls (< 30 m) with high slope gradients of ~4.6° (Förster et al. 2010; Kopf et al. 2008). A slope failure east of Cap Ferrat Ridge is termed Eastern slide (ES) and shows deep slide scars (80-120 m) and a lower slope gradient of ~3° (Förster et al. 2010; Kopf et al. 2008). The lithostratigraphic succession of Western Slide is characterized by homogenous, fine-grained hemipelagic clayey silt with some coarse-grained sand intervals (Kopf et al. 2008). Areas east of Cap Ferrat Ridge are not connected to major fluvial input of the Var system and receive only hemipelagic sediments (Klaucke et al. 2000). The sediments are generally composed of well-bioturbated, homogenous, fine-grained hemipelagic deposits (Kopf et al. 2008).

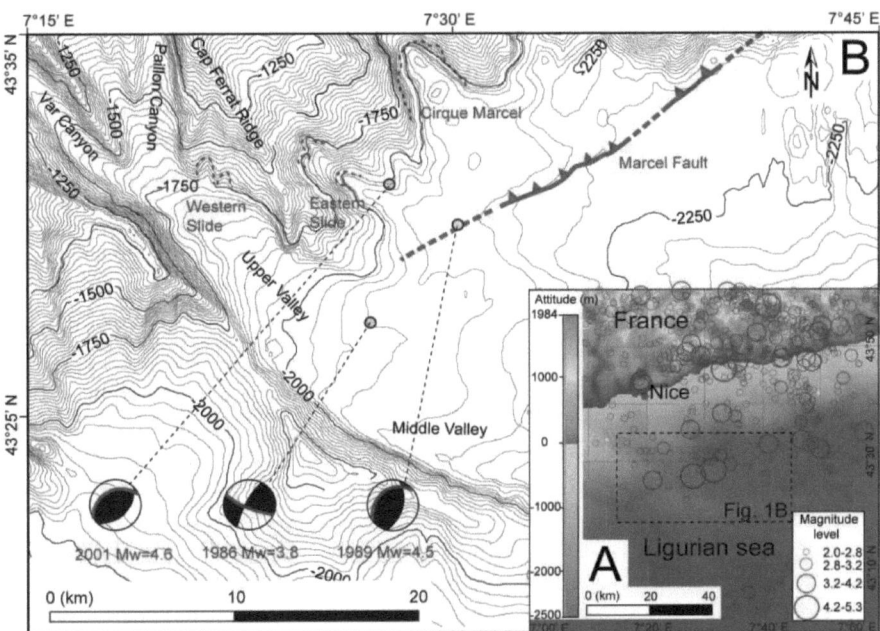

Fig. 1.8 *(A) Map showing the location of the study area, red circles indicate earthquake records of the Ligurian margin from 1980 to 2010 (catalogue from the Bureau Central Sismologique Français). (B) Bathymetric map of deeper slope of Ligurian margin with focal mechanisms of the moderate earthquakes associated with the Marcel Fault (taken from Larroque et al., 2012).*

1.4 Methods

Understanding submarine landslide preconditioning factors and triggering mechanisms requires integrated investigations that include seafloor surface morphology and undersurface sedimentary sequences by geophysical methods, sediment lithology by core recovery, and physical and geotechnical parameters of sediment by geotechnical methods. The necessity for multidisciplinary approaches to accurately recognize the submarine landslide processes arises from their societal threat.

1.4.1 Geophysical methods

The main devices used for geophysical investigation include bathymetric multibeam echo sounder (e.g., Kongsberg EM120), parametric sediment echo sounder (e.g., Atlas Parasound), and high-resolution multichannel seismic systems. The Kongsberg EM 120 bathymetric multibeam system operates at a frequency of 12 kHz to characterize the surface morphology of the working areas. It uses 191 beams with a maximal opening angle of 140°, corresponding to a swath width of 5.5 times the water depth (used in Cruise M78/3, Krastel et al. 2012). A 50-m grid spacing was chosen for the data presented in this thesis. Atlas Parasound system is a hull-mounted high-frequency sediment echo sounder used to image the sedimentary sequences in the sub-seafloor. It utilizes the so-called parametric effect to generate an operational signal of 4 kHz focused within a cone of 4° opening angle (used in Cruise M78/3, Krastel et al. 2012). Multichannel seismic profiles were acquired using a 1.71-1 GI-Gun served as source. A wavelet with main frequencies between 100 and 500 Hz was recorded by a 600 m-long, 96-channel streamer (used in Cruise M78/3, Krastel et al. 2012). The processing procedure of seismic profiles includes trace editing, setting up geometry, static corrections, velocity analysis, normal moveout corrections, bandpass frequency filtering stacking, and time migration. A common midpoint spacing of 10 m was applied for the data presented (for more details see Krastel et al. 2012).

1.4.2 Coring and sedimentological methods

Sediment cores were taken based on the acoustic results. Two coring techniques had been deployed in the study areas and allow sampling both the slide areas and the nearby non-failed areas. Gravity corer is the most basic of all the sediment core samplers because it is merely a tube that is weighted at the top. The corer penetrated under the influence of its own weigh using free-fall method (Fig. 1.9). Due to the simple design, the gravity corer can sample nearly any depth of water. Unfortunately, recovered samples tend to be shorter due to the restricted penetration depth. So the newly designed MeBo seafloor drill rig (Marum) is deployed to recover sediment core down to deeper below seafloor. MeBo corer is an electro-hydraulically driven robotic drill rig that is deployed on the seafloor and remotely operated from the research vessel (Fig. 1.10). Up to 70-m long core from soft sediments to hard rocks in water depths of up to 2 km can be drilled with MeBo using a wire-line coring technique (Freudenthal and Wefer 2009). Such deeper core may then provide a geotechnical characterization of slide plane material.

After recovery, the gravity cores were cut into segments of 1 m length. Then the segments were split in two halves: the archiving half for sedimentological description (macroscopically and using smear slide techniques; Rothwell 1989) and MSCL (GeoTek Multi Sensor core logger; see next section) measurements; the working half for geotechnical tests.

Grain size distribution analysis was performed using a Beckmann Coulter Counter LS 13320 particle size analyzer, which covers a size range from 0.04 μm to 2 mm. The percent of sand, silt and clay size grains in the samples is based on the classification scheme of Craig (2004).

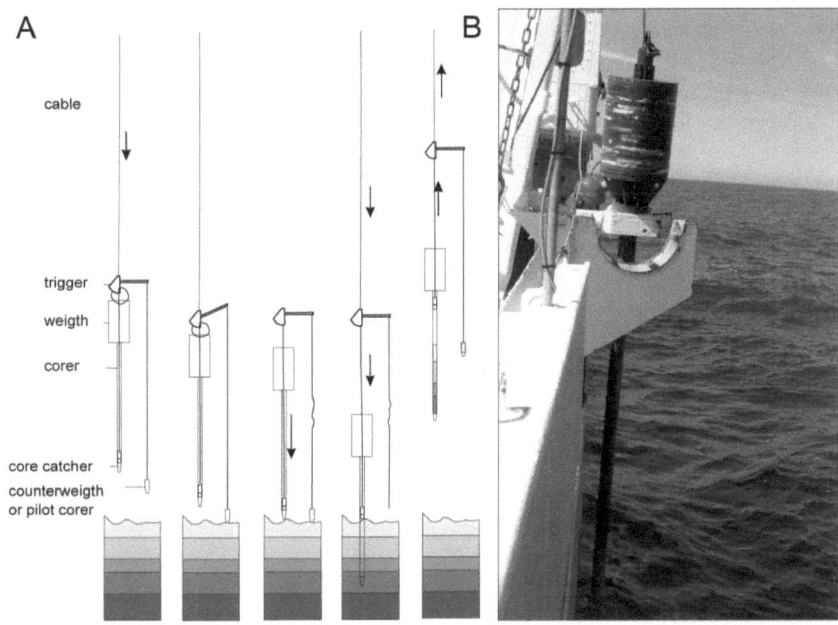

Fig. 1.9 *(A) Principle of core recovery with a gravity corer (modified after Mulder et al. 2011). (B) The picture of gravity corer.*

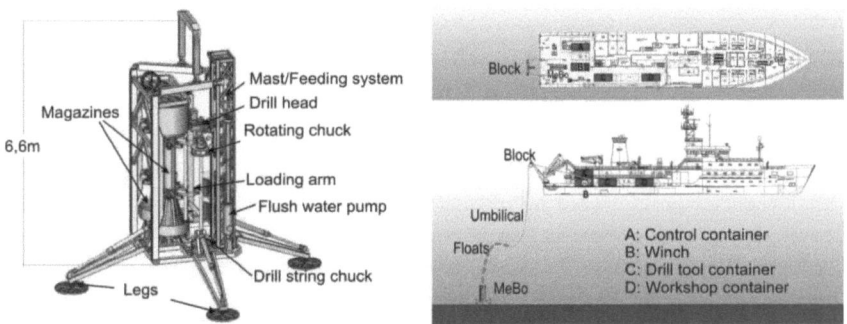

Fig. 1.10 *Schematic diagram showing the MeBo drill rig (left) and its deployment from a research vessel (right) (taken from http://www.marum.de/en/MARUM-MeBo_drill_rig.html).*

1.4.3 Geotechnical methods

MSCL analysis

Magnetic susceptibility was divided by the volume magnetization M (A/m) to the magnetizing field H (A/m). Magnetic susceptibility is a dimension physical quantity that represent the amount to which a material is magnetized by an external field. Magnetic susceptibility was measured at 2 cm intervals down core using a GeoTek Multi Sensor core logger (MSCL) (Blum 1997). Other physical parameters such as P-wave velocity, Gamma-ray density can also be obtained using MSCL (for more details see Cruise report M73/1; Kopf et al. 2008).

MAD analysis

Water content, the ratio of the mass of water to the mass of solids in the soil, was measured on discrete samples taken every 50 cm core depth immediately after core splitting, by weighing 10 cm^3 of sediment before and after drying in the oven at 105°C. The volume of a specimen can be measured by Pentapycnometry. Then the bulk density, grain density, porosity and void ratio can be determined using two-phase relationships assuming sediments are saturated (Craig 2004).

Atterberg limits

The Atterberg limits are a basic measure of the nature of a fine-grained soil and serve to mechanically distinguish between different types of silts and clays. The liquid limit (w_L) of samples was determined via the Casagrande apparatus (Casagrande 1932). The plastic limit (w_P) was determined by rolling a thread of soil without crumbling. The range over which the soil remains plastic is defined by the plasticity index (I_P):

$$I_P = w_L - w_P \tag{1.13}$$

The natural water content (w) of a soil relative to the liquid and plastic limits can be represented by means of the liquidity index (I_L):

$$I_L = \frac{w - w_P}{I_P} \tag{1.14}$$

Oedometer test

Consolidation test simulate how porosity evolves with effective stress under one-dimensional gravitational compaction caused by sedimentation. The transition from recompression to virgin compression behavior provides an estimate of the maximum in situ effective stress the sample has undergone (Casagrande 1936). The general sample preparation and testing procedures followed the guidelines set out by the ASTM 2435-04 standard (ASTM 2004b). Incremental loading consolidation test was conducted by an odometer (Fig. 1.11A). Before testing, the sample was trimmed to a height 1.48 cm and a diameter of 7.14 cm. A consolidation test was performed by applying 10 incremental loading steps from 4.9 kPa to 1962 kPa normal stress onto the sample and each load step was maintained for 24 h to allow complete pore pressure dissipation such that the applied vertical stress is equal to the vertical effective stress.

Plots of consolidation data show void ratio at the end of loads steps against the logarithm of the vertical effective stress (Fig. 1.11B). A pre-consolidated sample undergoes a recompression phase, a primary consolidation phase, and an unloading phase. For the different phases of the consolidation test, the recompression index (C_r), the compression index (C_c) and the expansion index (C_e) were determined respectively. Among them, the compression index (C_c) is the slope of a linear potion in the plot given by

$$C_c = \Delta e / \Delta \log(\sigma'_v) \tag{1.15}$$

The Casagrande method was used to estimate the value of the maximum past effective stress (σ'_{pc}) experienced by the sediment (i.e. its former maximum overburden), which is termed preconsolidation stress (Casagrande 1936). The stress history of sediments is often described using the overconsolidation ratio (OCR), which is a ratio of σ'_{pc} and in situ effective stress σ'_{vh} under hydrostatic conditions.

$$OCR = \sigma'_{pc} / \sigma'_{vh} \tag{1.16}$$

The vertical hydrostatic effective stress is calculated by integrating the sediment's bulk density from MAD or MSCL measurements and sample depth below the seafloor.

$$\sigma'_{vh} = \sum (\rho_b - \rho_w) \times \Delta z \times g \tag{1.17}$$

Where ρ_b is bulk density of the depth interval (g/cm^3), ρw is density of water (g/cm^3), z is depth interval (m), g is gravity acceleration (m/s^2). If OCR = 1, the sediment is considered to be normally consolidated. For OCR < 1, it is referred to as underconsolidated while OCR > 1 means that the sediment is overconsolidated. Overconsolidation typically results from sediment erosion which leads to overburden loading of the sediment experiencing at present lower than it has experienced. Consolidation tests can provide information on the degree of maximum past burial or how much previously overlying sedimentary column has been eroded from a section of sediment (Silva et al. 2001). The estimated maximum erosion depth (d_e) in meters can be calculated as follows:

$$d_e = (OCR \times d) - d \tag{1.18}$$

Where d is the sample depth in meters below the seafloor (mbsf). In most marine deposits, the uppermost sediments have strength attributed to cohesion higher than the pre-consolidation stress, and therefore are in a state of apparent overconsolidation resulting from weak interparticle bonds or bioturbation (Lee and Baraza 1999).

Oedometer test results also can reflect the disturbance of sediments. Two methods were used to quantitatively evaluate the amount of disturbance in samples from consolidation test (Fig. 1.11C). The first method was proposed by Silva (1974). Silva (1974) defined a disturbance index (I_D) as;

$$I_D = \Delta e / \Delta e_0 \tag{1.19}$$

Where Δe is the change in void ratio from the initial void ratio (e_0) to the void ratio corresponding to the laboratory compression line. Δe_0 is the change in void ratio from the laboratory compression line to the void ratio to the idealized remolded baseline. The categories of disturbance based on the values of I_D (Fig. 1.11C). The second method for assessing sample quality was proposed by Lunne et al. (1997). These authors suggested $\Delta e/e_0$ was more systematically influenced by sample disturbance. The different levels of disturbance were shown in Fig. 1.11C.

Furthermore, consolidation results also can be used to estimate the overpressure in sediments (more detail is found in section 1.1.2) and estimate permeability of sediments (more detail sees following part).

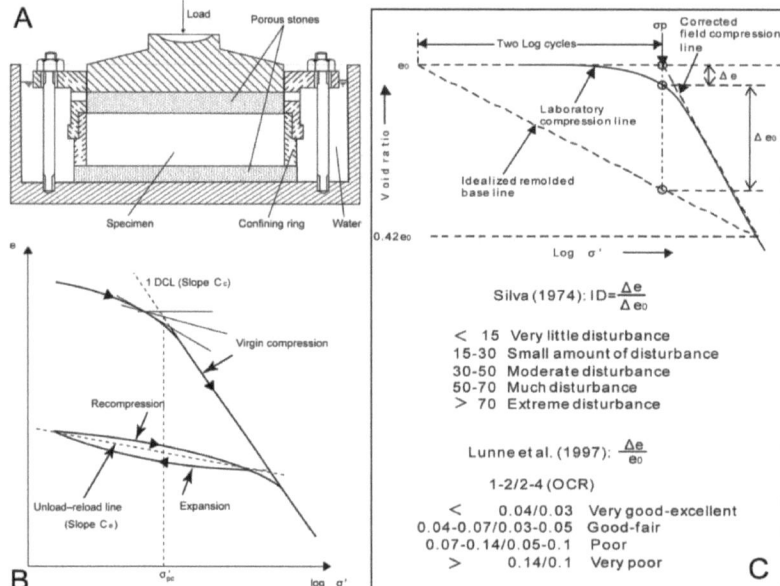

Fig. 1.11 (A) *Consolidation test arrangement (taken from Craig 2004). (B) Void ratio-effective stress relationship and determination of preconsolidation stress (σ'_{pc}). (C) Disturbance criteria proposed by Silva (1974) and Lunne et al. (1997).*

Permeability test

Permeability is an important property of porous sediments that controls fluid flow. The constant-flow permeability test was used to estimate vertical and horizontal hydraulic conductivity values for core samples from study areas. The constant-flow approach pumps fluid into and out the sample, and the resulting hydraulic gradient is measured (Fig. 1.12). Using specified flow rate Q (m^3/s) and the pressure difference across the sample for each flow rate (dP) monitored by the testing equipment, the hydraulic conductivity k (or coefficient of permeability) (m/s) was calculated for each sample using Darcy's Law,

$$Q = -k \times A \times (dh/dl) \qquad (1.20)$$

Where k is hydraulic conductivity (m/s), A is the area of the sample (m^2), and dl is the length of the sample (m), dh is the difference in head across the sample (m),

$$dh = \frac{dP}{\gamma_w \times g} \qquad (1.21)$$

Where γ_w represents unit weight of water (9.81 kN/m^3) and g is the gravitational constant (9.81 m/s^2).

The conductivity value was then converted to permeability K (m²) using the following equation:

$$K = \frac{k \times \mu}{\rho \times g} \quad (1.22)$$

Where μ is viscosity (0.0008Pa×s), ρ is density of water (1.02 g/cm³).
Permeability anisotropy r_k is commonly defined as the ratio of the hydraulic conductivity parallel to the bedding plane (i.e., horizontal flow), k_h, to the hydraulic conductivity in the direction perpendicular to the bedding plane (i.e., vertical flow), k_v.

$$r_k = k_h / k_v \quad (1.23)$$

The samples were trimmed to a height of about 7 cm and a diameter of 3.57 cm. ASTM 5084-03 standard was used as guidelines for general procedures (ASTM 2003). Different effective stress values up to 140 kPa were applied on the samples to monitor the change of the coefficient of permeability at different depth levels.
Through Terzaghi's theory of one-dimensional consolidation, the coefficient of permeability (k) can also be calculated by equation following using consolidation test results:

$$k = c_v \times m_v \times \gamma_w \quad (1.24)$$

Where m_v is the coefficient of volume compressibility (m²/MN), defined as the volume change per unit volume per unit increase in effective stress. If, for an increase in effective stress form σ'_0 to σ'_1, the void ratio decrease from e_0 to e_1, then

$$m_v = \frac{1}{1+e_0}\left(\frac{e_0 - e_1}{\sigma'_1 - \sigma'_0}\right) \quad (1.25)$$

c_v is the coefficient of consolidation (m²/year). Since k and mv are assumed as constants, c_v is constant during consolidation. The log time method (Casagrande method) was used to determine c_v (Craig 2004):

$$c_v = \frac{0.196 \times d^2}{t_{50}} \quad (1.26)$$

Where d is half of the average thickness of the specimen for the particular pressure increment, t_{50} is the time consuming when arriving at the half point for primary consolidation (more detail see Craig 2004).

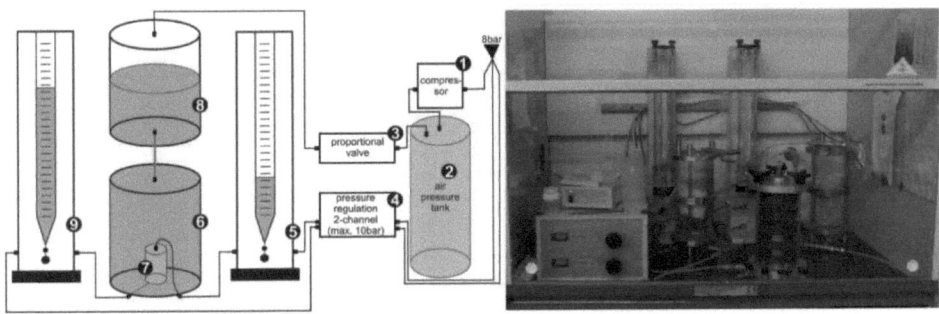

Fig. 1.12 Schematic diagram showing the high pressure constant-flow permeameter, 1: Air pressure booster, 2: Air pressure tank, 3: Air pressure regulating proportional valve, 4: Air pressure Regulator, 5: Tailwater burette, 6: Triaxial Test Cell, 7: Test Specimen inside flexible membrane, 8: Medium decollator, 9: Headwater burette (left) and picture of test apparatus (right).

Undrained and drained shear tests

The undrained shear strength (S_u) of the sediment was estimated using a Wykeham Farrance cone penetrometer at 10 cm intervals (Wood 1985) and a Mennerich Geotechnik (Germany) vane shear apparatus (rotation 90 °/min) at 50 cm intervals (Boyce 1977; Blum 1997). Remoulded undrained shear strength (S_{ur}) and strength sensitivity (S_u/S_{ur}) were determined from vane shear tests (Fig.1.13A). Vane shear test is one of the most versatile and widely used devices used for investigating undrained shear strength (S_u) and sensitivity of soft clays. The two independently measured undrained shear strength data sets are considered for first-order comparison of magnitude and used to identify characteristic strength versus depth trends. Since the test is very fast, unconsolidated undrained (UU) can be expected.

Consolidated drained (CD) shear strength is used to evaluate long-term slope stability. The cohesion (c') and angle of internal friction (ϕ') were derived from direct shear test on discrete undisturbed samples to complement the undrained shear-strength shipboard measurements (see above). The shear strength (τ) of a soil at a point on a particular plane was expressed based on Coulomb theory as a linear function of the effective normal stress at failure (σ') on the plane at the same point (Craig 2004):

$$\tau = c' + \sigma' \times \tan \phi' \tag{5}$$

The ASTM 3080-04 (ASTM 2004a) standard was used as guideline for the general procedures. Consolidated drained direct shear tests (CD) were performed using a displacement-controlled direct shear device built by Giesa GmbH (Germany) (Fig. 1.13B). The samples were trimmed to a height of 2 cm and a diameter of 5.05 cm. Vertical loads was applied to consolidate the sample for at least 24 h or until consolidation settlement ceased. After complete dissipation of the excess pore pressure, the shear test was performed. The shear rates are 0.002 mm/min for clay and 0.008 mm/min for silt. From the plot of the shear stress versus the horizontal displacement, the peak shear stress (shear strength τ_f) was obtained for a specific vertical confining stress (σ'). The experiment was run several times (at least three times) for various σ' (from 50 kPa to 300 kPa), and a plot of the σ' versus τ_f for each test was derived. Then the Mohr-Coulomb failure envelope curve can be drawn and c' and ϕ' can be determined.

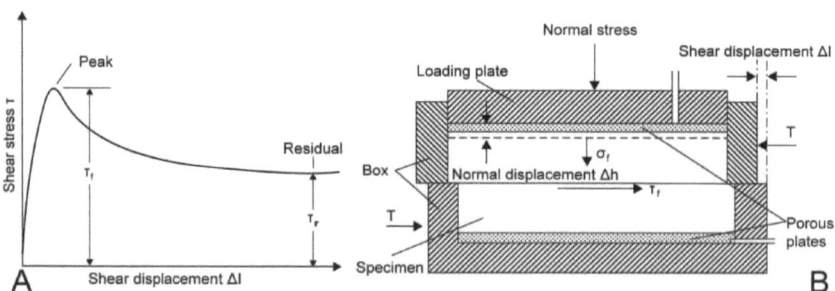

Fig. 1.13 (A) *Shear displacement-shear stress relationship and determinations of peak shear strength and residual shear strength. (B) Schematic diagram of direct shear apparatus.*

1.4 Outline of the thesis

This thesis contributes to understand the preconditioning factors and trigger mechanisms of submarine landslides in different continental margin settings: the passive margin (Northern Argentina and Uruguay Margin) and tectonically active margins (Mediterranean continental margins: the Gela Basin and the Ligurian margin). To achieve this aim, three manuscripts that make up the core chapters of this thesis are described briefly in the following:

Chapter 2, submitted to *Landslides*, deals with small-scale landslide features on the open slope and canyon area off the Uruguayan and Argentine margin using geophysical, sedimentological and geotechnical data. Infinite slope stability analysis and the pseudostatic earthquake approach allow us to determine the most likely condition for slope failure initiation and influence of earthquake on slope instability. Geotechnical investigations of sediments from undeformed sediments upslope of the landslide allow a reconstruction of the sedimentological processes acting on the slope and discussion of their influence on physical and geotechnical properties that affect slope failure initiation and slide modes.

Chapter 3, accepted by *6th International Symposium on Submarine Mass Movements and Their Consequences* and to be published in the Springer Conference Proceedings book, presents a case study of a relatively small scale, 8 kyr old landslide named Northern Twin Slide (NTS) on the slope of the Gela Basin in the Sicily Channel (central Mediterranean). Sedimentological and geotechnical data of a MeBo core taken from undeformed upslope of headwall of NTS allows for assessment current state of slope stability, the plausible controls for past failure, and potential causes for future failures.

Chapter 4, submitted to *the World Landslide Forum 3*, presents geotechnical measurements and infinite slope stability analysis of small-scale landslides on the deeper slope of the Ligurian margin offshore Southern France. Sedimentological and geotechnical properties of sediments from the undeformed, headwall and deformed deposit areas of two distinct landslides allow us to assess the influences of sedimentological processes and earthquakes on slope failure initiation and style.

References

ASTM (2003) Standard test method for measurement of hydraulic conductivity of saturated porous materials using a flexible wall permeameter (Standard D5084-03). Annual Book of ASTM Standards: Philadelphia (Am. Soc. Testing and Mater.). ASTM International, West Conshohocken, United States.

ASTM (2004a) Standard test method for direct Shear test of soils under consolidated drained conditions. ASTM International, Pennsylvania, United States.

ASTM (2004b) Standard test methods for one-dimensional consolidation properties of soils using incremental loading ASTM International, West Conshohocken, United States.

Argnani A (1990) The strait of sicily rift zone: Foreland deformation related to the evolution of a back-arc basin. Journal of Geodynamics 12 (2-4): 311-331.

Bethoux N, Ouillon G, Nicolas M (1998) The instrumental seismicity of the western Alps: spatio-temporal patterns analysed with the wavelet transform. Geophysical Journal International 135 (1): 177-194.

Bindi D, Pacor F, Luzi L, Puglia R, Massa M, Ameri G, Paolucci R (2011) Ground motion prediction equations derived from the Italian strong motion database. Bulletin of Earthquake Engineering 9 (6): 1899-1920.

Blum P (1997) Physical Properties Handbook: A guide to the shipboard measurement of physical properties of deep-sea cores. ODP Technical Notes 26: 118.

Bondevik S, Løvholt F, Harbitz C, Mangerud J, Dawson A, Inge Svendsen J (2005) The Storegga Slide tsunami—comparing field observations with numerical simulations. Marine and Petroleum Geology 22 (1–2): 195-208.

Boyce RRE (1977) Deep Sea Drilling Project procedures for shear strength measurement of clayey sediment using modified Wykeham Farrance laboratory vane apparatus. Initial Reports DSDP, vol 36. U.S. Government Printing Office, Washington.

Bozzano G, Violante R, Cerredo M (2011) Middle slope contourite deposits and associated sedimentary facies off NE Argentina. Geo-Marine Letters 31 (5-6): 495-507.

Bryn P, Berg K, Stoker MS, Haflidason H, Solheim A (2005) Contourites and their relevance for mass wasting along the Mid-Norwegian Margin. Marine and Petroleum Geology 22 (1-2): 85-96.

Butler RWH, Grasso M, Lamanna F (1992) Origin and deformation of the Neogene–Recent Maghrebian foredeep at the Gela Nappe, SE Sicily. Journal of the Geological Society 149 (4): 547-556.

Campbell KW, Bozorgnia Y (2008) NGA ground motion model for the geometric mean horizontal component of PGA, PGV, PGD and 5% damped linear elastic response spectra for periods ranging from 0.01 to 10 s. Earthquake Spectra 24: 139-171.

Canals M, Lastras G, Urgeles R, Casamor JL, Mienert J, Cattaneo A, De Batist M, Haflidason H, Imbo Y, Laberg JS, Locat J, Long D, Longva O, Masson DG, Sultan N, Trincardi F, Bryn P (2004) Slope failure dynamics and impacts from seafloor and shallow sub-seafloor geophysical data: case studies from the COSTA project. Marine Geology 213 (1-4): 9-72.

Casagrande A (1932) Research on the Atterberg limits of soils. Public roads 13 (8): 121-136.

Casagrande A The determination of the pre-consolidation load and its practical significance. In: Proceedings of the 1st International Conference of Soil Mechanics and Foundation Engineering,

1936. pp 60-64.

Cochonat P, Bourillet JF, Savoye B, Dodd L (1993) Geotechnical characteristics and instability of submarine slope sediments, the nice slope (N-W Mediterranean Sea). Marine Georesources & Geotechnology 11 (2): 131-151.

Coleman JM (1988) Dynamic changes and processes in the Mississippi River delta. Geological Society of America Bulletin 100 (7): 999-1015.

Courboulex F, Deschamps A, Cattaneo M, Costi F, Déverchère J, Virieux J, Augliera P, Lanza V, Spallarossa D (1998) Source study and tectonic implications of the 1995 Ventimiglia (border of Italy and France) earthquake (ML=4.7). Tectonophysics 290 (3-4): 245-257.

Craig RF (2004) Craig's soil mechanics. Taylor & Francis.

Dewey JF, Helman ML, Knott SD, Turco E, Hutton DHW (1989) Kinematics of the western Mediterranean. Geological Society, London, Special Publications 45 (1): 265-283.

Dugan B, Flemings PB (2000) Overpressure and fluid flow in the New Jersey continental slope: Implications for slope failure and cold seeps. Science 289 (5477): 288-291.

Dugan B, Flemings PB (2002) Fluid flow and stability of the US continental slope offshore New Jersey from the Pleistocene to the present. Geofluids 2 (2): 137-146.

Dugan B, Sheahan TC (2012) Offshore sediment overpressures of passive margins: Mechanisms, measurement, and models. Rev Geophys 50 (3): RG3001.

Ercilla G, Casas D (2012) Submarine Mass Movements: Sedimentary Characterization and Controlling Factors. Earth Sciences.

Ewing M, Lonardi AG (1971) Sediment transport and distribution in the Argentine Basin. 5. Sedimentary structure of the Argentine margin, basin, and related provinces. Physics and Chemistry of The Earth 8: 123-251.

Fine IV, Rabinovich AB, Bornhold BD, Thomson RE, Kulikov EA (2005) The Grand Banks landslide-generated tsunami of November 18, 1929: preliminary analysis and numerical modeling. Marine Geology 215 (1-2): 45-57.

Flemings PB, Long H, Dugan B, Germaine J, John CM, Behrmann JH, Sawyer D, Scientists IE (2008) Pore pressure penetrometers document high overpressure near the seafloor where multiple submarine landslides have occurred on the continental slope, offshore Louisiana, Gulf of Mexico. Earth and Planetary Science Letters 269 (3-4): 309-325.

Flood RD, Shor AN (1988) Mud waves in the Argentine Basin and their relationship to regional bottom circulation patterns. Deep Sea Research Part A Oceanographic Research Papers 35 (6): 943-971.

Förster A, Spieß V, Kopf AJ, Dennielou B (2010) Mass Wasting Dynamics at the Deeper Slope of the Ligurian Margin (Southern France). In: Mosher DC, Shipp C, Moscardelli L et al. (eds) Submarine Mass Movements and Their Consequences. Andvances in Natural and Technological Hazard Research Springer, Dordrecht, Heidelberg, London, New York, pp 66-77.

Freudenthal T, Wefer G Shallow drilling in the deep sea: The sea floor drill rig MeBo. In: OCEANS 2009 - EUROPE, 11-14 May 2009 2009. pp 1-4.

Franke D, Neben S, Ladage S, Schreckenberger B, Hinz K (2007) Margin segmentation and volcano-tectonic architecture along the volcanic margin off Argentina/Uruguay, South Atlantic. Marine Geology 244 (1-4): 46-67.

Gardiner W, Grasso M, Sedgeley D (1995) Plio-pleistocene fault movement as evidence for mega-block kinematics within the Hyblean-Malta Plateau, Central Mediterranean. Journal of Geodynamics 19 (1): 35-51.

Garming JFL, Bleil U, Riedinger N (2005) Alteration of magnetic mineralogy at the sulfate-methane transition: Analysis of sediments from the Argentine continental slope. Physics of the Earth and Planetary Interiors 151 (3-4): 290-308.

Gattacceca J, Deino A, Rizzo R, Jones DS, Henry B, Beaudoin B, Vadeboin F (2007) Miocene rotation of Sardinia: New paleomagnetic and geochronological constraints and geodynamic implications. Earth and Planetary Science Letters 258 (3-4): 359-377.

Giberto DA, Bremec CS, Acha EM, Mianzan H (2004) Large-scale spatial patterns of benthic assemblages in the SW Atlantic: the Río de la Plata estuary and adjacent shelf waters. Estuarine, Coastal and Shelf Science 61 (1): 1-13.

Gibson R (1958) The progress of consolidation in a clay layer increasing in thickness with time. Geotechnique 8 (4): 171-182.

Hampton MA, Lee HJ, Locat J (1996) Submarine landslides. Reviews of Geophysics 34 (1): 33-59.

Hance JJ (2003) Submarine Slope Stability. The University of Texas at Austin, Austin.

Harders R, Kutterolf S, Hensen C, Moerz T, Brueckmann W (2010) Tephra layers: A controlling factor on submarine translational sliding? Geochemistry, Geophysics, Geosystems 11 (5): Q05S23.

Heezen BC, Ewing M (1952) Turbidity currents and submarine slumps, and the 1929 Grand Banks Earthquake. American Journal of Science 250: 849-873.

Hinz K, Neben S, Schreckenberger B, Roeser HA, Block M, Souza KGd, Meyer H (1999) The Argentine continental margin north of 48°S: sedimentary successions, volcanic activity during breakup. Marine and Petroleum Geology 16 (1): 1-25.

Hjelstuen BO, Sejrup HP, Haflidason H, Berg K, Bryn P (2004) Neogene and Quaternary depositional environments on the Norwegian continental margin, 62° N-68° N. Marine Geology 213 (1-4): 257-276.

Hsü KJ, W.B.F.Ryan, M.B.Cita (1973) Late Miocene desiccation of the Mediterranean. Nature 242 (5395): 240-244.

Keller J, Ryan WBF, Ninkovich D, Altherr R (1978) Explosive volcanic activity in the Mediterranean over the past 200,000 yr as recorded in deep-sea sediments. Geological Society of America Bulletin 89 (4): 591-604.

Klaucke I, Cochonat P (1999) Analysis of past seafloor failures on the continental slope off Nice (SE France). Geo-Marine Letters 19 (4): 245-253.

Klaucke I, Savoye B, Cochonat P (2000) Patterns and processes of sediment dispersal on the continental slope off Nice, SE France. Marine Geology 162 (2-4): 405-422.

Kopf A, Participants C (2008) Report and Preliminary Results of Meteor Cruise M 73/1: LIMA-LAMO (Ligurian Margin Landslide Measurements & Observatory). vol 264. Berichte Fachbereich Geowissenschaften, Universität Bremen.

Krastel S, Lehr J, Winkelmann D, Schwenk T, Preu B, Strasser M, Wynn RB, Georgiopoulou A, Hanebuth T Mass wasting along Atlantic continental margins: a comparison between NW-Africa and the de la Plata River region (northern Argentina and Uruguay). In: Krastel et al. (ed) Submarine Mass Movements and Their Consequences Advances in Natural and Technological Hazard Research, Kiel, accepted.

Krastel S, Wefer G, Hanebuth TJ, Antobreh A, Freudenthal T, Preu B, Schwenk T, Strasser M, Violante R, Winkelmann D (2011) Sediment dynamics and geohazards off Uruguay and the de la Plata River region (northern Argentina and Uruguay). Geo-Marine Letters 31 (4): 271-283.

Krastel S, Wefer G, participants c (2012) Report and preliminary results of RV METEOR Cruise

M78/3. Sediment transport off Uruguay and Argentina: from the shelf to the deep sea; 19.05. 2009-06.07. 2009, Montevideo (Uruguay)-Montevideo (Uruguay). vol No. 285. Berichte, Fachbereich Geowissenschaften, Universität Bremen.

Krastel S, Wynn RB, Hanebuth TJJ, Henrich R, Holz C, Meggers H, Kuhlmann H, Georgiopoulou A, Schulz HD (2006) Mapping of seabed morphology and shallow sediment structure of the Mauritania continental margin, Northwest Africa: some implications for geohazard potential. Norwegian Journal of Geology 86 (3): 163-176.

Kvalstad TJ, Nadim F, Kaynia AM, Mokkelbost KH, Bryn P (2005) Soil conditions and slope stability in the Ormen Lange area. Marine and Petroleum Geology 22 (1-2): 299-310.

Labbé M, Donnadieu C, Daubord C, Hébert H (2012) Refined numerical modeling of the 1979 tsunami in Nice (French Riviera): Comparison with coastal data. J Geophys Res 117 (F1): F01008.

Laberg J, Dahlgren T, Vorren T, Haflidason H, Bryn P (2001) Seismic analyses of Cenozoic contourite drift development in the Northern Norwegian Sea. Marine Geophysical Researches 22 (5-6): 401-416.

Laberg JS, Camerlenghi A (2008) The Significance of Contourites for Submarine Slope Stability. In: Rebesco M, Camerlenghi A (eds) Developments in Sedimentology, vol Volume 60. Elsevier, pp 537-556.

Laberg JS, Vorren TO, Mienert J, Evans D, Lindberg B, Ottesen D, Kenyon NH, Henriksen S (2002) Late Quaternary palaeoenvironment and chronology in the Trænadjupet Slide area offshore Norway. Marine Geology 188 (1–2): 35-60.

Laberg JS, Vorren TO, Mienert J, Haflidason H, Bryn P, Lien R (2003) Preconditions Leading to the Holocene Trænadjupet Slide Offshore Norway. In: Locat J, Mienert J, Boisvert L (eds) Submarine Mass Movements and Their Consequences, vol 19. Advances in Natural and Technological Hazards Research. Springer Netherlands, pp 247-254.

Larroque C, Scotti O, Ioualalen M (2012) Reappraisal of the 1887 Ligurian earthquake (western Mediterranean) from macroseismicity, active tectonics and tsunami modelling. Geophysical Journal International 190 (1): 87-104.

Lee HJ, (2005) Undersea landslides: extent and significance in the Pacific Ocean, an update. Natural Hazards and Earth System Sciences 5 (6): 877-892.

Lee H, Baraza J (1999) Geotechnical characteristics and slope stability in the Gulf of Cadiz. Marine Geology 155 (1-2): 173-190.

Lee HJ, Edwards BD (1986) Regional Method to Assess Offshore Slope Stability. Journal of Geotechnical Engineering 112 (5): 489-509.

Lee HJ, Locat J, Desgagnés P, Parsons JD, McAdoo BG, Orange D, Puig P, Wong FL, Dartnell P, Boulanger E (2007) Submarine mass movements on continental margins. In: Nittrouer CA, Austin JA, Field ME, Kravitz JH, Syvitski JPM, Wiberg PL (eds) Continental Margin Sedimentation: From Sediment Transport to Sequence Stratigraphy. pp 213-273.

Lee HJ, Orzech K, Locat J, Boulanger E, Konrad J Seismic strengthening, a conditioning factor influencing submarine landslide development. In: 57th Canadian Geotechnical Conference, 2004.

Lee HJ, Schwab WC, Edwards BD, Kayen RE (1991) Quantitative controls on submarine slope failure morphology. Marine Geotechnology 10 (1-2): 143-157.

Lermusiaux PFJ, Robinson AR (2001) Features of dominant mesoscale variability, circulation patterns and dynamics in the Strait of Sicily. Deep Sea Research Part I: Oceanographic Research Papers 48 (9): 1953-1997.

Leroueil S (2001) Natural slopes and cuts: movement and failure mechanisms. Géotechnique 51 (3): 197-243.

Leynaud D, Mienert J, Vanneste M (2009) Submarine mass movements on glaciated and non-glaciated European continental margins: A review of triggering mechanisms and preconditions to failure. Marine and Petroleum Geology 26 (5): 618-632.

Locat J, Lee H (2009) Submarine Mass Movements and Their Consequences: An Overview. In: Sassa K, Canuti P (eds) Landslides-Disaster Risk Reduction. Springer-Verlag Berlin Heidelberg, pp 115-142.

Locat J, Lee HJ (2002) Submarine landslides: advances and challenges. Canadian Geotechnical Journal 39 (1): 193-212.

Lunne T, Berre T, Strandvik S Sample disturbance effects in soft low plastic Norwegian clay. In: Symposium on Recent Developments in Soil and Pavement Mechanics, 1997.

Madof AS, Christie-Blick N, Anders MH (2009) Stratigraphic controls on a salt-withdrawal intraslope minibasin, north-central Green Canyon, Gulf of Mexico: Implications for misinterpreting sea level change. AAPG Bulletin 93 (4): 535-561.

Martinsen JO, Bakken B (1990) Extensional and compressional zones in slumps and slides in the Namurian of County Clare, Ireland. Journal of the Geological Society 147 (1): 153-164.

Martinsen OJ (2005) Deep-water sedimentary systems of Arctic and North Atlantic margins: An introduction. Norwegian Journal of Geology 85: 161-166.

Martorelli E, Petroni G, Chiocci F (2011) Contourites offshore Pantelleria Island (Sicily Channel, Mediterranean Sea): depositional, erosional and biogenic elements. Geo-Marine Letters 31 (5-6): 481-493.

Masson DG, Harbitz CB, Wynn RB, Pedersen G, Løvholt F (2006) Submarine landslides: processes, triggers and hazard prediction. Philosophical Transactions of the Royal Society A: Mathematical, Physical and Engineering Sciences 364 (1845): 2009-2039.

Masson DG, Wynn RB, Talling PJ (2010) Large Landslides on Passive Continental Margins: Processes, Hypotheses and Outstanding Questions. In: Mosher DC, Shipp C, Moscardelli L et al. (eds) Submarine Mass Movements and Their Consequences. Andvances in Natural and Technological Hazard Research Springer, pp 153-165.

Matano RP, Palma ED, Piola AR (2010) The influence of the Brazil and Malvinas Currents on the Southwestern Atlantic Shelf circulation. Ocean Science 6 (4): 983-995.

McHugh CMG, Damuth JE, Mountain GS (2002) Cenozoic mass-transport facies and their correlation with relative sea-level change, New Jersey continental margin. Marine Geology 184 (3-4): 295-334.

Mienert J, Berndt C, Laberg JS, Vorren TO (2003) Slope Instability of Continental Margins. In: Wefer G, Billett D, Hebbeln D (eds) Ocean Margin Systems. Springer Verlag, New York, pp 179-193.

Migeon S, Cattaneo A, Hassoun V, Larroque C, Corradi N, Fanucci F, Dano A, Mercier de Lepinay B, Sage F, Gorini C (2011) Morphology, distribution and origin of recent submarine landslides of the Ligurian Margin (North-western Mediterranean): some insights into geohazard assessment. Marine Geophysical Research 32 (1-2): 225-243.

Minisini D, Trincardi F (2009) Frequent failure of the continental slope: The Gela Basin (Sicily Channel). Journal of Geophysical Research 114 (F03014).

Minisini D, Trincardi F, Asioli A, Canu M, Foglini F (2007) Morphologic variability of exposed mass-transport deposits on the eastern slope of Gela Basin (Sicily channel). Basin Research 19:

217-240.

Morgenstern N (1967) Submarine slumping and the initiation of turbidity currents. Marine geotechnique: 189-220.

Mosher DC, Moscardelli L, Baxter CDP, Urgeles R, Shipp RC, Chaytor JD, Lee HJ, Mosher DC, Moscardelli L, Shipp RC, Chaytor JD, Baxter CDP, Lee HJ, Urgeles R (2010) Submarine Mass Movements and Their Consequences. In: Submarine Mass Movements and Their Consequences, vol 28. Advances in Natural and Technological Hazards Research. Springer Netherlands, pp 1-8.

Mulder T (2011) Gravity Processes and Deposits on Continental Slope, Rise and Abyssal Plains. Deep-Sea Sediments 63: 25-148.

Mulder T, Cochonat P (1996) Classification of offshore mass movements. Journal of Sedimentary Research 66 (1): 43-57.

Mulder T, Hüneke H, Van Loon A (2011) Progress in Deep-Sea Sedimentology. Deep-Sea Sediments 63:1-24.

Mulder T, Savoye B, Syvitski JPM (1997) Numerical modelling of a mid-sized gravity flow: The 1979 Nice turbidity current (dynamics, processes, sediment budget and seafloor impact). Sedimentology 44 (2): 305-326.

Nadim F (2012) Risk Assessment for Earthquake-Induced Submarine Slides. Submarine Mass Movements and Their Consequences: 15-27.

Nelson CH, Escutia C, Damuth JE, Twichell DC (2011) Interplay of mass-transport and turbidtite-system deposits in different activie tectonic and passive continental margin settings. In: Mass-Transport Deposits in Deep water Settings, vol SEPM Special Publication. SEPM (Society for Sedimentary Geology), pp 39-66.

Parsons JD, Friedrichs CT, Traykovski PA, Mohrig D, Imran J, Syvitski JPM, Parker G, Puig P, Buttles JL, García MH (2007) The Mechanics of Marine Sediment Gravity Flows. Continental Margin Sedimentation. Blackwell Publishing Ltd.

Piola AR, Matano RP (2001) Brazil And Falklands (malvinas) Currents. In: Editor-in-Chief: John HS (ed) Encyclopedia of Ocean Sciences. Academic Press, Oxford, pp 340-349.

Piola AR, Matano RP, Palma ED, Möller OO, Jr., Campos EJD (2005) The influence of the Plata River discharge on the western South Atlantic shelf. Geophysical Research Letters 32 (1): L01603.

Piper DJW, Cochonat P, Morrison ML (1999) The sequence of events around the epicentre of the 1929 Grand Banks earthquake: initiation of debris flows and turbidity current inferred from sidescan sonar. Sedimentology 46 (1): 79-97.

Preu B, Hernández-Molina FJ, Violante R, Piola AR, Paterlini CM, Schwenk T, Voigt I, Krastel S, Spiess V (2013) Morphosedimentary and hydrographic features of the northern Argentine margin: The interplay between erosive, depositional and gravitational processes and its conceptual implications. Deep Sea Research Part I: Oceanographic Research Papers 75 (0): 157-174.

Preu B, Schwenk T, Hernández-Molina FJ, Violante R, Paterlini M, Krastel S, Tomasini J, Spieß V (2012) Sedimentary growth pattern on the northern Argentine slope: The impact of North Atlantic Deep Water on southern hemisphere slope architecture. Marine Geology 329–331 (0): 113-125.

Rehault J-P, Bethoux N (1984) Earthquake relocation in the Ligurian Sea (western Mediterranean): Geological interpretation. Marine Geology 55 (3–4): 429-445.

Reuther C (1987) Extensional tectonics within the central Mediterranean segment of the Afro-European zone of convergence. Mem Soc Geol It 38: 69-80.

Rothwell RG (1989) Minerals and mineraloids in marine sediments: an optical identification guide.

Elsevier Applied Science. London.

Sahal A, Lemahieu A (2011) The 1979 nice airport tsunami: mapping of the flood in Antibes. Natural Hazards 56 (3): 833-840.

Savoye B, Piper DJW (1991) The Messinian event on the margin of the Mediterranean Sea in the Nice area, southern France. Marine Geology 97 (3–4): 279-304.

Savoye B, Piper DJW, Droz L (1993) Plio-Pleistocene evolution of the Var deep-sea fan off the French Riviera. Marine and Petroleum Geology 10 (6): 550-571.

Sawyer DE, Flemings PB, Dugan B, Germaine JT (2009) Retrogressive failures recorded in mass transport deposits in the Ursa Basin, Northern Gulf of Mexico. J Geophys Res 114 (B10): B10102.

Seed HB (1979) Considerations in the earthquake-resistance design of earth and rockfill dams. Geotechnique 29 (3): 215-263.

Seed HB, Idriss IM (1971) Simplified procedure for evaluation soil liquefaction potential Soil Mechanics and Foundation Engineering SM 9: 1249-1273.

Seed HB, Seed RB, Schlosser F, Blondeau F, Juran I (1988) The landslide at the Port of Nice on October 16, 1979, vol 88. vol 10. Earthquake Engineering Research Center, University of California.

Silva AJ (1974) Marine geomechnics: overview and projections. In: Inderbitzen AL (ed) Deep sea sediments.

Silva AJ, LaRosa P, Brausse M, Baxter CD, Bryant WR (2001) Stress states of marine sediments in plateau and basin slope areas of the northwestern Gulf of Mexico. Offshore Technology Conference.

Skempton AW, Hutchinson J (1969) Stability of natural slopes and embankment foundations. Paper presented at the Soil mechanical and foundation conference, Mexico.

Solheim A, Bryn P, Sejrup HP, Mienert J, Berg K (2005) Ormen Lange-an integrated study for the safe development of a deep-water gas field within the Storegga Slide Complex, NE Atlantic continental margin; executive summary. Marine and Petroleum Geology 22 (1-2): 1-9.

Solheim A, Forsberg CF, Yang S, Kvalstad TJ (2007) The Role of Geological Setting and Depositional History in Offshore Slope Instability. Offshore Technology Conference.

Sosa AB (1998) Sismicidad y sismotectónica en Uruguay. Física de la tierra (10): 167-186.

Stegmann S, Strasser M, Kopf A, Anselmetti FS (2007) Geotechnical in situ characterisation of subaquatic slopes: The role of pore pressure transients versus frictional strength in landslide initiation. Geophysical Research Letter: L07607.

Stigall J, Dugan B (2010) Overpressure and earthquake initiated slope failure in the Ursa region, northern Gulf of Mexico. J Geophys Res 115 (B4): B04101.

Strasser M, Hilbe M, Anselmetti F (2011) Mapping basin-wide subaquatic slope failure susceptibility as a tool to assess regional seismic and tsunami hazards. Marine Geophysical Research 32 (1-2): 331-347.

Strozyk F (2009) Submarine landslides in active margin environments slope stability vs. neotectonic activity on the northeastern margin of Crete, eastern Mediterranean. University of Bremen, Bremen.

Sultan N, Cochonat P, Canals M, Cattaneo A, Dennielou B, Haflidason H, Laberg JS, Long D, Mienert J, Trincardi F, Urgeles R, Vorren TO, Wilson C (2004) Triggering mechanisms of slope instability processes and sediment failures on continental margins: a geotechnical approach. Marine Geology 213 (1-4): 291-321.

ten Brink US, Barkan R, Andrews BD, Chaytor JD (2009) Size distributions and failure initiation of submarine and subaerial landslides. Earth and Planetary Science Letters 287 (1-2): 31-42.

Terzaghi K, Peck RB, Mesri G (1996) Soil mechanics in engineering practice. Wiley.

Tinti S, Maramai A, Graziani L (2004) The New Catalogue of Italian Tsunamis. Natural Hazards 33 (3): 439-465.

Trincardi F, Argnani A (1990) Gela submarine slide: A major basin-wide event in the plio-quaternary foredeep of Sicily. Geo-Marine Letters 10 (1): 13-21.

Verdicchio G, Trincardi F (2008) Mediterranean shelf-edge muddy contourites: examples from the Gela and South Adriatic basins. Geo-Marine Letters 28 (3): 137-151.

Violante RA, Paterlini CM, Costa IP, Hernández-Molina FJ, Segovia LM, Cavallotto JL, Marcolini S, Bozzano G, Laprida C, García Chapori N, Bickert T, Spieß V (2010) Sismoestratigrafía y evolución geomorfológica del talud continental adyacente al litoral del este bonaerense, Argentina. Latin American journal of sedimentology and basin analysis 17: 33-62.

Wilson CK, Long D, Bulat J (2004) The morphology, setting and processes of the Afen Slide. Marine Geology 213 (1–4): 149-167.

Wood DM (1985) Some fall-cone tests. Géotechnique 38: 64-68.

Yamada Y, Kawamura K, Ikehara K, Ogawa Y, Urgeles R, Mosher D, Chaytor J, Strasser M (2012) Submarine Mass Movements and Their Consequences. In: Yamada Y, Kawamura K, Ikehara K et al. (eds) Submarine Mass Movements and Their Consequences, vol 31. Advances in Natural and Technological Hazards Research. Springer Netherlands, pp 1-12.

2 Geotechnical characteristics and slope stability along the Uruguayan and northern Argentine margin

Fei Ai[1,], Benedict Preu[1], Till Hanebuth[1], Sebastian Krastel[2], Michael Strasser[3] and Achim Kopf[1]*

[1]MARUM-Center for Marine Environmental Sciences, and Faculty of Geosciences University of Bremen, Leobener Straße, 28359 Bremen, Germany

[2] Institute for Geosciences, Kiel University, Germany

[3]Geological Institute, ETH Zurich, Sonneggstrasse 5, 8092 Zurich, Switzerland

[*]Corresponding author: aifei@uni-bremen.de

(Submitted to Landslides)

Abstract

Submarine mass movements are common along the Uruguayan and northern Argentine slope, primarily because of high fluvial discharge by Rio de la Plata River and strong bottom current forces within the Brazil-Malvinas Current Confluence (BMC) zone. This study combines sedimentological, physical and geotechnical results of core samples to quantitatively assess slope stability for two distinct study areas: mass movements dominated area along the Uruguayan slope called Northern Slide area (NS) and canyons dominated area off the Rio de la Plata River mouth of Northern Argentine slope called Southern Canyon area (SC). NS mainly consists of clayey silt with interbedded sand layers with wide changes of physical and geotechnical properties from surficial (0-3 m) to deeper sediments (> 3 m): bulk density (1.5-2.1 g/cm^3), water content (20-95%), void ratio (0.6-3.0) and undrained shear strength (5-200 kPa from 0 to 16 m below seafloor (mbsf)). SC mainly contains silty sand with high bulk density (1.7-2.4 g/cm^3), low water content (20-40%), low void ratio (0.6-1.2) and low undrained shear strength (5-20 kPa from 0 to 20 mbsf). Oedometer tests of both sites show overconsolidated (overconsolidation ratio, OCR: 1.5-12.7) near the seafloor and underconsolidated (OCR: 0.13-0.2) at depths of 20-30 mbsf. Direct shear tests indicate that NS materials have a lower angle of internal friction (30.3-34.3°) compared to those of SC (36.9-41.3°).

Slope stability analyses in different scenarios (undrained static, drained static, undrained earthquake and drained earthquake) indicate both sites are stable under static conditions. SC is vulnerable in the undrained static case (factor of safety (FS) = 1-2 at slope angle 3-5°). Overpressure is unlikely to trigger slope failure at both sites in the drained static case. Under earthquake loading, SC requires low peak ground acceleration (PGA: ~0.043 g) to experience slope failure in the undrained earthquake case, while moderate PGA (0.125-0.13 g) is needed at NS to trigger slope failure in the drained earthquake case. We propose that NS is sufficiently stable and is unlikely to experience repeated small-scale slope failures under the current conditions, but may experience unstable conditions if external triggers (e.g., earthquakes) are strong enough to trigger slope failure. In contrast, low stability of SC's steep slopes is reflected by repeated small-scale slope failures both during static conditions and certainly during seismic events.

Keywords Slope stability, Submarine landslide, Geotechnical characteristics, Uruguayan and *northern Argentine margin*

2.1 Introduction

Submarine mass movements are widespread on submarine slopes which play an important role in transporting sediments across the continental slope to the deep basin, as well as potential endangering seabed infrastructure (Masson et al. 2006). For most of submarine landslides, the exact mechanisms of slope instability are not well understood certainly. Preconditioning factors of sediments, such as overpressure leading to decrease frictional resistance and the existence of mechanical weak layers, promote slopes susceptible to instability (Flemings et al. 2008; Harders et al. 2010). Triggering mechanisms such as earthquake forces imposing on the slope sediments and rapid changes in slope geometry by undercutting or oversteepening ultimately trigger slope failure (Sultan et al. 2004; Leynaud et al. 2009).

The continental margin off Uruguay and northern Argentina is located within the Brazil-Malvinas Current Confluence (BMC) zone (35-38° S). This area is also characterized by high fluvial discharge by the Rio de la Plata River and strong bottom current forces by different water masses. As a result, this area is an excellent site to study interaction between gravitational downslope and contour current-driven alongslope sediment transports along with their influences on seabed morphology (Krastel et al. 2011). Mass movements are common on open slope and canyon areas off the Uruguayan and Argentine margin (Lonardi and Ewing 1971; Klaus and Ledbetter 1988; Krastel et al. 2011). Our study area actually comprises two survey subareas: a northern one called Northern Slide area (NS) northeast off the Rio de la Plata River mouth between 35.7-36.4° S on the Uruguayan slope, a southern one called Southern Canyon area (SC) close to Mal de la Plata canyon between 37.4-37.7° S on the northern Argentine slope. For NS, Krastel et al. (2011) used geophysical and sedimentological methods to study the morphology owing to mass movements. Henkel et al. (2011) applied geochemical and geotechnical methods on surficial mass transport deposits (MTDs) to reconstruct recent submarine landslide scenarios. For SC, Preu et al. (2012, 2013) used seismic data to explain the evolution of contourite terraces. Bozzano et al. (2011) described sedimentary facies to reconstruct contourite deposition. These studies provide a strong background for further quantitative slope stability analyses.

Here we present physical and geotechnical results of core samples taken from undisturbed slopes of NS and SC adjacent to scarps in order to simulate slope stability under various conditions. The main objective of this study is to investigate how geometrical and geotechnical parameters affect seabed morphology and predict future failure modes of slopes.

2.2 Regional geological, morphologic and oceanographic settings

The Uruguayan and northern Argentine margin is considered an extensive volcanic passive margin which formed during the opening of the South Atlantic in the early Cretaceous (Hinz et al. 1999). The Argentine margin has been subdivided into four tectonic segments separated by transfer fracture zones (Franke et al. 2007). The study area is located in the northernmost segment separated by the Salado transfer zone (Fig. 2.1A). Current tectonic activity in the study area is characterized by active subsidence, which resulted in some intraplate seismic activities aligned along the Salado transfer zone

(Sosa 1998). Recent documented earthquakes occurred in the years of 1849, 1888, and 1988 A.D. (for locations of epicenters, see Fig. 2.1A, Sosa 1998). The epicenter and magnitude of the 1988 earthquake is not conclusively defined. Seismological observatory of the University of Brazil showed the epicenter located 36.5° S, 53.5° W, +/- 100 km with the regional magnitude of 3.9, whereas, NEIC (National Earthquake Information Center) recorded the epicenter of 36.27° S, 52.73° W with a body wave magnitude of 5.1 m_b (Sosa 1998). Here we choose the seismic parameters as the epicenter of 36.27° S, 52.73° W with a body wave magnitude of 5.2 m_b following Assumpção (1998).

Based on the seabed morphology, our study area is separated into two distinct subareas: northern mass movements dominated area (NS) on the Uruguayan slope and southern canyons dominated area (SC) on the northern Argentine slope. Krastel et al. (2011) refer to NS as the "northern slide area" and "drift-and-scarp area". NS is characterized by a smooth topography and gentle slopes (1-3°) typical of margins where deposition prevails over erosion (Ewing and Lonardi 1971). Several scarps (Fig. 2.1B, C) and mass transport deposits (MTDs) (Figs. 2.2A, A', B, B') at 1200-2800 m were found in NS (detail descriptions see Krastel et al. 2011). SC is characterized by the flat Ewing terrace, which is dissected by canyons at ~1400 m (e.g., Mar del Plata Canyon and Querendi Canyon, Fig 2.1D). The slope below the Ewing terrace is steep (3-7°), suggesting that erosion has been the main process shaping the slope of SC (Ewing and Lonardi 1971). No clear MTDs were found in the sedimentary sequences upper slope and flanks of the Querendi Canyon (Fig. 2.2C, D). However, MTDs were imaged in the thalweg of the Querendi Canyon, which suggests headward erosion is common in this canyon (Fig. 9 of Krastel et al. 2011)

At present, sedimentation in the study area is strongly influenced by high fluvial discharge of the Rio de la Plata River and strong bottom current forces. The sediments of the study area are generally divided into a clayey silt NS part and sand SC part (Huppertz 2011). The boundary between the two environments is located at the position of the Brazil Malvinas confluence (BMC, ~37° S). North of the BMC, high quantities of terrigenous sediments (~80×10^6 ton/yr) discharged by Rio de la Plata River, containing 75% coarse to medium silt, 15% fine to very fine silt and 10% clay (Giberto et al. 2004), are swept northwards by alongshore currents and deposited at the Uruguayan continental shelf and slope (Piola et al. 2005); sediment transport is dominated by the southward flow of North Atlantic Deep Water (NADW) at the Uruguayan slope in water depths between 2000-4000 m. South of the BMC, coarse fluvial sediments are trapped with the estuary and occasionally carried directly down slope by turbidity current (Garming et al. 2005); most of the surficial sediments at the northern Argentine slope have been carried and reworked northward by water masses in different depths including the Antarctic Intermediate Water (AAIW, ~500-1000 m), the Circumpolar Deep Water (~1000-3500 m) and the Antarctic Bottom Water (AABW, >3500 m) (more details, see Piola and Matano 2001).

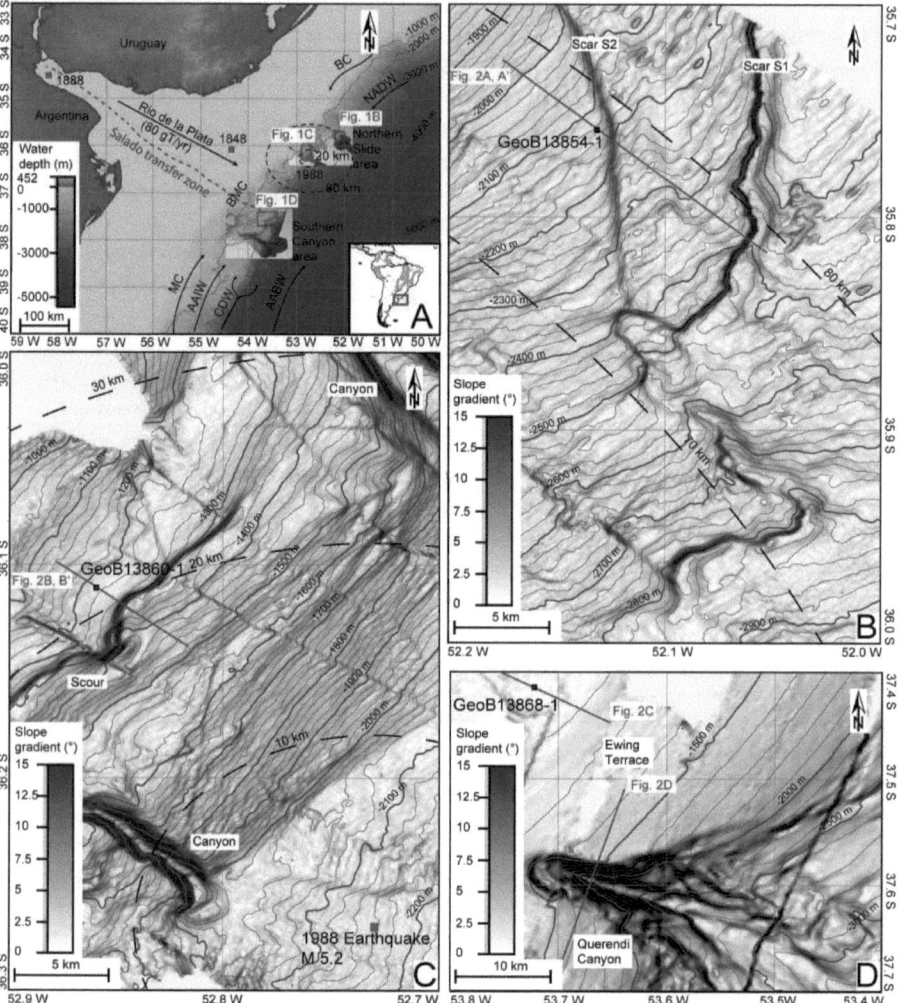

Fig. 2.1 *(A) Map showing the location of the study area and the oceanographic setting along the Uruguayan and northern Argentine margin. Red dots indicate the epicenters of the 1848, 1888 and 1988 earthquakes. MC: Malvinas Current, AAIW: Antarctic Intermediate Water, CDW: Circum Polar Deep Water, AABW: Antarctic Bottom Water, BC: Brazil Confluence, and NADW: North Atlantic Deep Water. (B) and (C) Bathymetric contour maps of NS. Red lines indicate the position of Parasound and seismic profiles. Black dots indicate core locations. Black dashed lines indicate the distances to the epicenter of the 1988 earthquake. D) Bathymetric map of SC. Modified after Krastel et al. (2011).*

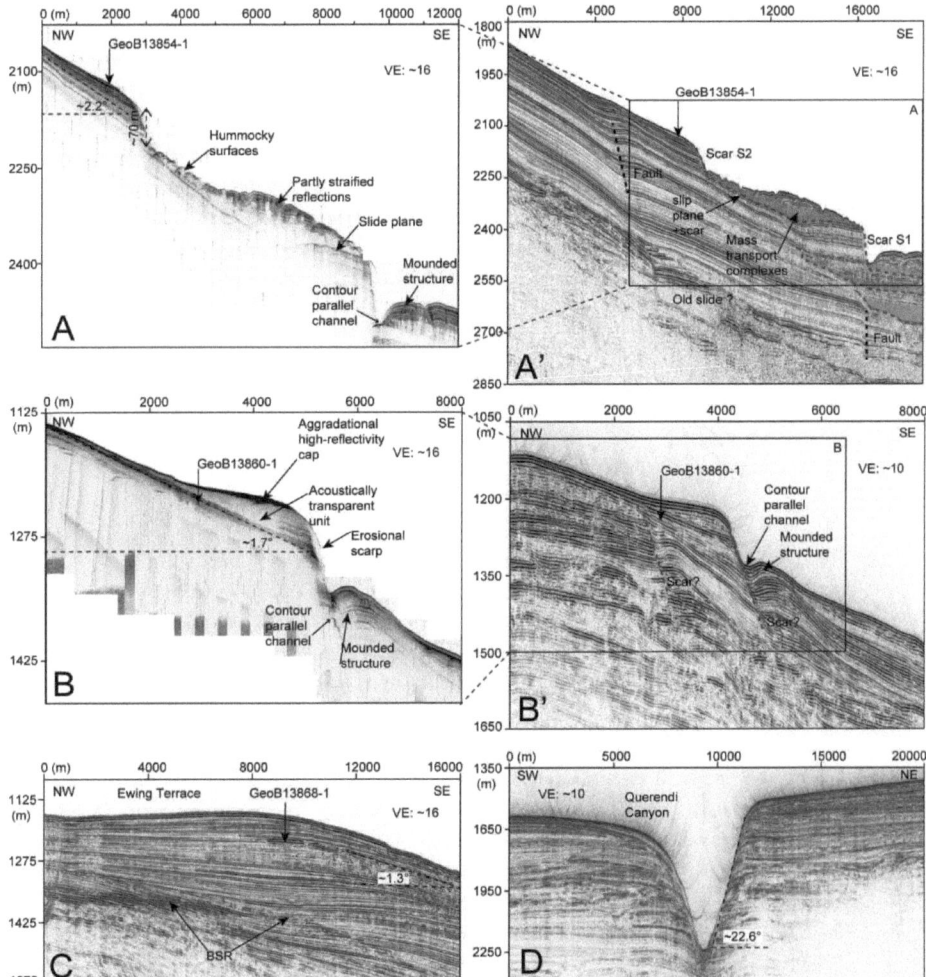

Fig. 2.2 (A) and (B) Parasound profiles of the NS area; (A') and (B') Seismic profiles of NS; (C) and (D) Parasound profiles of SC. Modified after Krastel et al. (2011)

2.3 Material and methods

The database comprising sediment gravity cores and MeBo (German for seafloor drill rig) cores was gained during Meteor Cruise M78/3 that took place from 19 May to 6 July 2009 offshore northern Argentina and Uruguay. This study focuses on one gravity core (GeoB13854-1: drill depth of 5.52 m at ~2120 mbsf) and one MeBo core (GeoB13860-1: drill depth of 35.6 m at ~1220 mbsf with 82% recovery) recovered in undisturbed upperslope sediments adjacent to the scars of NS, and one MeBo core (GeoB13868-1: drill depth of 21.5 m at ~1140 mbsf with 34% recovery) recovered in the upperslope sediment of SC (core locations see Fig. 2.1). The sedimentological and physical properties of sediments were obtained by shipboard tests, and further geotechnical parameters were gained in the laboratory tests.

2.3.1 Shipboard tests

Visual core decription was carried out shortly after core recovery on the split core. GeoTeK Multi Sensor Core Logger (MSCL) was used to measure bulk density (by Gamma ray attenuation) and magnetic susceptibility on the archive halves at 2 cm intervals. Discrete samples at 50 cm intervals were taken on the working halves to measure water content (moisture content), bulk density, void ratio using oven drying method and pycnometre (Moisture and Density, MAD; Blum 1997). The undrained shear strength (S_u) of the sediment was estimated using a Wykeham Farrance cone penetrometer at 10 cm intervals (Wood 1985) and a Mennerich Geotechnik (Germany) vane shear apparatus (rotation 90 °/min) at 50 cm intervals (Boyce 1977; Blum 1997). Remoulded undrained shear strength (S_{ur}) and strength sensitivity (S_u/S_{ur}) were determined from vane shear tests. The two independently measured undrained shear strength data sets are considered for first-order comparison of magnitude and used to identify characteristic strength versus depth trends.

2.3.2 Laboratory tests

Grain size distribution analysis was performed using a Beckmann Coulter Counter LS 13320 particle size analyzer, which covers a size range from 0.04 μm to 2 mm. The percent of sand, silt and clay size grains in the samples is based on the classification scheme of Craig (2004). Atterberg limits including liquid limit (w_L), plastic limit (w_P) and plasticity index ($I_P = w_L-w_P$) served to distinguish between different types of silts and clays. Liquid limit was determined with a Casagrande apparatus and plastic limit was determined using the rolling thread method (Casagrande 1932). For evaluating the consolidation history of the sediments, one dimensional incremental loading oedometer tests were conducted on whole-round samples taken at different subbottom depths. Specimens of 1.5 cm height and 5 cm in diameter were trimmed and incrementally subjected to normal loads from 4.9 kPa to 1962 kPa (ASTM 2004b). Consolidated-drained direct shear strength tests were performed using a displacement-controlled direct shear apparatus built by Giesa GmbH (Germany) in order to obtain the drained shear strength of the sediments. Specimens of 2 cm height and 5 cm in diameter were consolidated at a specified normal load for at least 24 h. After complete dissipation of the excess pore pressure, the specimens were sheared with shear rate of 0.002 mm/min for clay and 0.008 mm/min for silt and sand (ASTM 2004a).

2.3.3 Overpressure estimation

Effective stress is an important parameter for slope stability analysis. Overpressure impacts effective stress as seen in Terzaghi et al. (1996). Overpressure (Δu) is defined as fluid pressure (u) in excess of hydrostatic equilibrium (u_0). Terzaghi's effective stress relationship follows:

$$\sigma'_v = \sigma_v - u = \sigma_v - (u_0 + \Delta u) = (\rho_b - \rho_w)gz - \Delta u = (\gamma - \gamma_w)z - \Delta u = \gamma'z - \Delta u \tag{1}$$

Where σ'_v is vertical effective stress, σ_v is total overburden stress, ρ_b is bulk density, ρ_w is water density, γ is unit weight of the bulk sample, γ_w is unit weight of water, γ' is buoyant weight, z is overburden depth, and g is gravity acceleration. Since insitu pore pressure of deep sea sediment is difficult to measure, two methods were used to estimate overpressure in this study. Preconsolidation stress (σ'_{pc}) interpreted from oedometer tests are used to estimate overpressure (Casagrande 1936):

$$\Delta u = \sigma'_{vh} - \sigma'_{pc} \tag{2}$$

Where σ'_{vh} is vertical effective stress for hydrostatic conditions (σ'_{vh} = γ'z). Overpressure due to sedimentation can be evaluated with Gibson's (1958) one dimensional solution under the assumption that a constant sedimentation rate and no flow at underlying strata occurs. The modeled overpressure is controlled by Gibson's time factor (T_g) (Flemings et al. 2008):

$$T_g = \frac{m^2 t}{c_v} \tag{3}$$

Where m is sedimentation rate, t is time, and c_v is coefficient of consolidation ($c_v = k/(m_v \gamma_w)$), the latter of which depends on coefficient of permeability (k) and coefficient of volume compressibility (m_v), both being obtained from oedometer tests.

2.3.4 Slope stability assessment

According to the low average slope angles (~2°) of submarine slope and the low ratio between failure depth and spatial extent for submarine landslide along passive continental margin, the infinite slope stability is assumed to be appropriate using to calculate the factor of safety (FS). The factor of safety determines whether a given slope is stable (FS > 1) or susceptible to failure (FS ≤ 1). Slope failure occurs when the failure-inducing stresses acting on the slope exceed the failure-resisting strength of the sediment (Hampton et al. 1996). The shear strength of sediments depends on the conditions and time of drainage during shear. It is essential to consider long-term factors such as overpressure induced by sedimentation (drained condition) and short-term factors (undrained condition) such as forces induced by earthquakes. Slope stability was evaluated for four different scenarios:
(1) Static undrained conditions can be affected by rapid change in slope geometry or fluctuation of pore pressure. The factor of safety calculation after Morgenstern (1967) and Løseth (1999) follows:

$$FS = \frac{S_u}{\gamma' z \sin\theta \cos\theta} \quad \text{(Undrained static)} \tag{4}$$

Where θ is slope angle (also the assumed angle for slip surface).
(2) Static drained conditions respond to long-term steady state pore pressure (Dugan and Flemings 2002):

$$FS = \frac{c' + \gamma' z (\cos^2\theta - \lambda^*) \tan\phi'}{\gamma' z \sin\theta \cos\theta} \quad \text{(Drained static)} \tag{5}$$

Where c' is cohesion, ϕ' is angle of internal friction (both gained from drained direct shear tests), and λ* is overpressure ratio ($\lambda^* = \Delta u/\sigma'_{vh}$).

(3) Earthquake undrained condition uses pseudostatic analysis for a simplified evaluation of the seismic factor of safety of a slope. The earthquake force is represented by a horizontal force and a pseudostatic seismic coefficient (k_e). The pseudostatic acceleration (a) is k_e times the gravitational acceleration g (a = k_eg), which is assumed to be applied over a time period long enough for the induced shear stress to be considered being constant (Hampton et al 1996). The undrained pseudostatic factor of safety is given by the following expression (ten Brink et al. 2009):

$$FS = \frac{S_u}{\gamma'z\left[\sin\theta\cos\theta + k_e\left(\gamma/\gamma'\right)\cos^2\theta\right]} \quad \text{(Undrained earthquake)} \quad (6)$$

(4) Earthquake drained conditions only include pre-earthquake pore pressure (not considering the overpressure developed during seismic shaking) under the assumption that shear strength does not decrease during seismic shaking:

$$FS = \frac{c' + \gamma'\left(\cos^2\theta - \lambda^*\right)\tan\phi}{\gamma'z\left[\sin\theta\cos\theta + k_e\left(\gamma/\gamma'\right)\cos^2\theta\right]} \quad \text{(Drained earthquake)} \quad (7)$$

2.3.5 Prediction of peak ground acceleration (PGA)

The critical pseudostatic acceleration (a_c) is the earthquake acceleration at which earthquake induced stress just equals the shear strength (FS = 1 of Equations 6 and 7). Critical pseudostatic acceleration as the average equivalent uniform shear stress imposed by seismic shaking represents ~65% of the effective seismic peak ground acceleration (PGA = a_c/65%) (Seed and Idriss 1971; Seed 1979; Strasser et al. 2011). The median ground motion of peak ground acceleration was estimated using an empirical seismic attenuation relationship by Campbell and Bozorgnia (2008) in this study. The absolute value of PGA depends on magnitude, source distance, style of faulting of the earthquake, hanging-wall, site response and basin response. Here we only take into account the well-known parameters of magnitude and source distance for PGA determination using equation following:

$$\ln(PGA) = \begin{cases} (c_0 + c_1 M) + \left[(c_4 + c_5 M)\ln\left(\sqrt{R_{RUP}^2 + c_6^2}\right)\right]\cdots\cdots\cdots\cdots & 5.5 \\ (c_0 + c_1 M) + c_2(M - 5.5) + \left[(c_4 + c_5 M)\ln\left(\sqrt{R_{RUP}^2 + c_6^2}\right)\right]\cdots\cdots & M \leq 6.5 \\ (c_0 + c_1 M) + c_2(M - 5.5) + c_3(M - 6.5) + \left[(c_4 + c_5 M)\ln\left(\sqrt{R_{RUP}^2 + c_6^2}\right)\right]\cdots & 5 \end{cases} \quad (8)$$

Where M is the earthquake magnitude, R_{RUP} is the epicentral distance and c_{0-6} are empirical coefficients: c_0 = -1.715, c_1 = 0.5, c_2 = -0.53, c_3 = -0.262, c_4 = -2.118, c_5 = 0.17, c_6 = 5.6 (Campbell and Bozorgnia 2008).

2.4 Results

2.4.1 Physical and geotechnical properties of sediments

One gravity core GeoB13854-1 comes from the upper slope of the scar S2 of NS at a water depth of ~2120 m (Fig. 2.2B). Two MeBo cores were also used: Core GeoB13860-1 is located upslope of a scour ~60 km southwest of the scar S2 of NS at water depths of ~1220 m (Fig. 2.2C) and core GeoB13868-1 is located at the nearly flat portion of the middle continental slope (the Ewing terrace) ~20 km north of the Querendi Canyon of SC at water depths of ~1140 m (Fig. 2.2C). Sediment lithofacies were defined on the basis of visual core description (see Fig. 2.3) and quantitative parameters such as magnetic susceptibility (see Fig. 2.3), grain size distribution (see Fig. 2.4A) and Atterberg limits (see Fig. 2.4B). Visual core description shows that sediments of NS (cores GeoB13854-1 and GeoB13860-1) mainly consist of fine-grained materials (silt and clay) with interbedded sand layers, while the sediments of SC (GeoB13868-1) consist of very fine sand to medium sand (Fig. 2.3). Magnetic susceptibility provides a first order estimate of ferromagnetic mineral abundance in sediments and is sensitive to grain size variations, with higher grain size resulting in higher magnetic susceptibility of our cores (Fig. 2.3) (Bozzano et al. 2011). Grain size distribution and Atterberg limits from whole round samples give a more refined classification for sediments (Fig. 2.4). The representative grain size of sediments from NS is clayey silt (~20% clay, 10-20% sand) to sandy silt (core GeoB13854-1 at 4.71 m) with intermediate to high plasticity. The representative grain size of sediments from SC is silty sand (2-8% clay, 18-38% silt).

The key physical properties of sediments from NS and SC are compared in Figure 2.3. Physical properties such as bulk density, water content, and void ratio have close association with grain size distribution. The bulk density values obtained from MAD are slightly lower than the values obtained from MSCL. Silty sediments from NS show lower bulk density (1.5-2.1 g/cm^3) comparing to sandy sediments from SC (1.7-2.4 g/cm^3). Silty sediments from NS show high water content (65-95%) at surficial depth (GeoB13854-1 at 5 m) and lower water content (~20%) at deep depth (GeoB13860-1 at 35 m). Sandy sediments (GeoB13868-1) from SC show low water content between 20 and 50%. Void ratio has similar trends as water content showing highest void ratios (1.6-3.0) in core GeoB13854-1 and lowest void ratios (0.6-1.2) in core GeoB13868-1.

The values of the undisturbed and remoulded shear strength are presented in Figure 2.3. Undrained shear strengths from vane shear tests are lower than the values from cone penetrometer measurements at the same depth. Sediments of core GeoB13854-1 have high undrained strength (~20 kPa) near the top of the core and gradually increase to ~30 kPa at 5 m core depth. Coarse-grained sediments in the top 1.5 m of core GeoB13860-1 have low shear strength (0-20 kPa) and increase rapidly up to ~200 kPa at 16 m bsf (below sea floor). Undrained shear tests were not carried out below 16 mbsf because the sediments were found stiff to hard. The strength sensitivity of the sediments from NS obtained from vane shear test is low (1.0-2.7). Vane shear test were not conducted on the sandy sediments from SC. The values of shear strength obtained from cone penetrometer are in range of 0-30 kPa. The ratio of the undrained shear strength to vertical effective stress (S_u/σ_v) gives an indication of the consolidation state of sediment. Typical values of S_u/σ_v for normally consolidated sediments range between 0.2-0.4 (Locat and Lee 2002). It is indicated from the stiffness of the sediments at a given depth that materials from NS are highly overconsolidated, while the sediments from SC are underconsolidated to normally consolidated.

Results of oedometer test are presented in Table 2.1 and Figure 2.5A as plots of void ratio versus the logarithm of applied vertical effective stress. Samples from core GeoB13854-1 of NS show a large amount of compression (decrease in void ratio from 2 to 0.6), while the samples from core GeoB13868-1 of SC show a small amount of compression (decrease in void ratio from 1 to 0.6). The preconsolidation stress, which is the maximum effective stress the sediments have ever been subjected to, was determined by the classical graphic method of Casagrande (1936). The overconsolidation ratio, OCR (ratio of the preconsolidation stress to the present overburden effective stress) has been calculated to range between 12.73 and 0.14 for sediments from NS and between 1.49 and 0.13 for deposits from SC. The compression index (C_c), a measure of the compressibility of the sediment, shows an intermediate compressibility (C_c = 0.30-0.39) for surficial 5 mbsf sediments (core GeoB13854-1) and low compressibility (C_c = 0.12-0.26) for sediments down to 35 mbsf (core 13860-1) in the NS area, which is dominated by silty material. In contrast, very low compressibility (C_c = 0.014-0.039) is found for the sandy sediments from area SC. Values of coefficient of consolidation (c_v), a measure of the rate at which sediment consolidates, are relative low (c_v = $1.1e^{-8}$-$2.2e^{-7}$ m^2/s) for NS materials and higher (c_v = $6.0e^{-8}$-$2.9e^{-6}$ m^2/s) for SC materials. The result of vertical coefficient of permeability (k), which was back-calculated from c_v, is presented as plots of coefficient of permeability versus void ratio (Fig. 2.5B). Results lie in the range of $3.6e^{-10}$-$2.3e^{-9}$ (m/s) for sediments from NS and $2.6e^{-10}$-$3.0e^{-9}$ (m/s) for SC deposits.

Overpressure results estimated from preconsolidation stress appear to show overpressure ratios of 0.8-0.86 in NS and ~0.87 in SC, suggesting that non-equilibrium consolidation occurs in both areas. Using Gibson's analytical solution for consolidation, the degree of overpressure controlled by Gibson's time factor (Eq. 3) for both study areas is shown in Table 2.2. Though previous studies show high sediment rate (0.8-1.8 m/kyr) in recent deposits of NS (Henkel et al. 2011) and high sedimentation rate (up to 1m/kyr) at the thalweg of the canyon of SC during the Holocene (Voigt 2013), a reliable long-term sedimentation rate for study area is missing. Here, we assume average sedimentation rate of 0.1 m/kyr, time span for sedimentation is 500 kyr (estimation the overpressure at 50 mbsf) of both sites. Average coefficients of consolidation were obtained by oedometer tests. The results of overpressure ratio estimated by contour plots of Gibson's time factor with depth an overpressure show overpressure ratio range of 0.5-0.6 in NS and ~0.3 in SC.

Drained direct shear test results (Tab. 2.3) are presented as plots of shear stress versus horizontal displacement (Fig. 2.6A), and shear strength versus effective normal stress to construct the Mohr-Coulomb failure envelope (Fig. 2.6B). The stress-displacement curves of the sediments from NS are characterized by smooth peaks without a later drop of shear stress, while the sediments from SC are characterized by clear peaks with further drop of shear stress. Values of the effective cohesion intercept (c') range from 1.5-20.1 kPa in the sediments from NS and 12.9 kPa in the sediments from SC. Lower values (30.3-34.3°) of angle of internal friction (ϕ') are found in the fine-grained sediments from NS, while higher values (36.9-41.3°) are observed in coarse-grained sediments from SC.

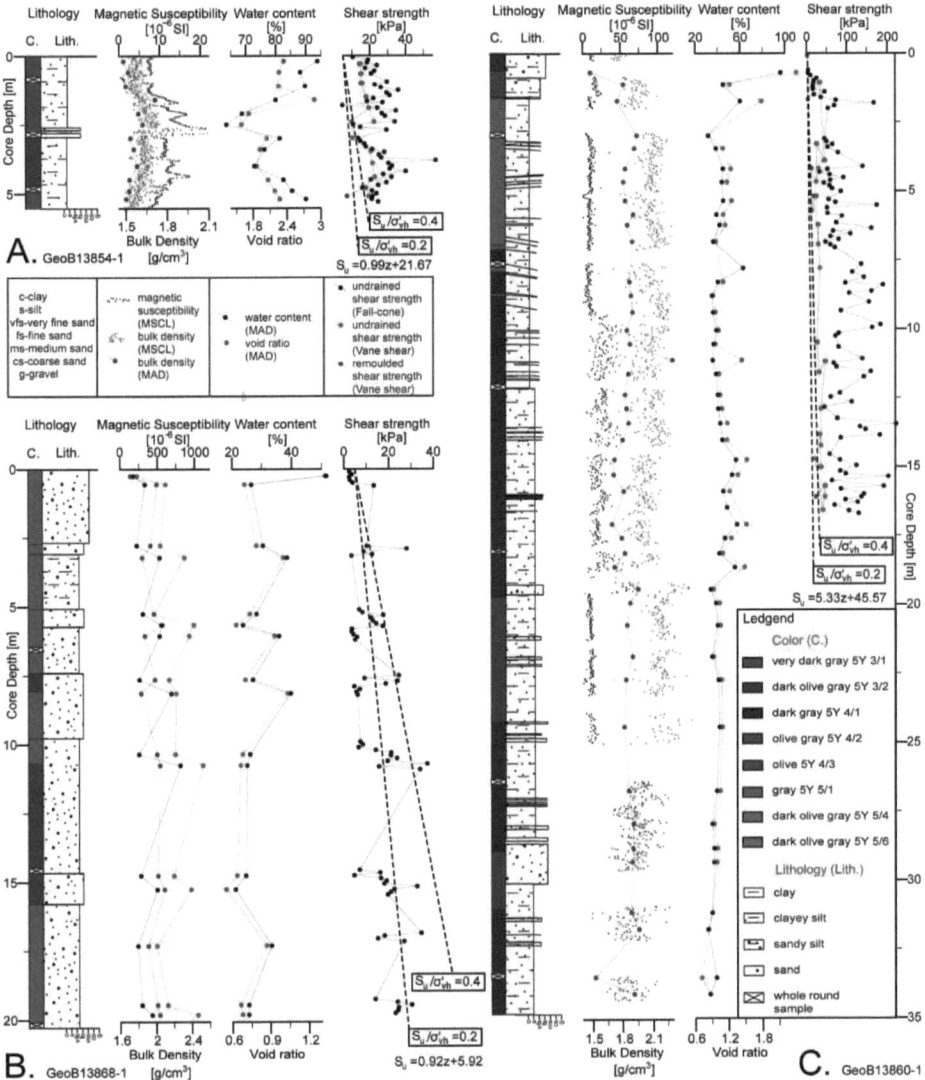

Fig. 2.3 *Physical and geotechnical properties of the sediments from NS (A) GeoB13854-1 and (B) GeoB13860-1 and from SC (C) GeoB13868-1. The dashed lines in the undrained shear strength plot are ratios for S_u/σ'_{vh} which show trends of underconsolidation, normal consolidation, and overconsolidation (≤ 0.2, 0.2-0.4, and ≥ 0.4, respectively), typical for fine-grained marine sediments (Locat and Lee 2002). Magnetic susceptibility and bulk density of GeoB13860-1 measured with the GeoTek MSCL are presented as mean average value per recovered core section, because the drilling process was highly destructive in the sandy material and fractionation/settling of heavy minerals towards the bottom of each segment was a result of fluidization during coring.*

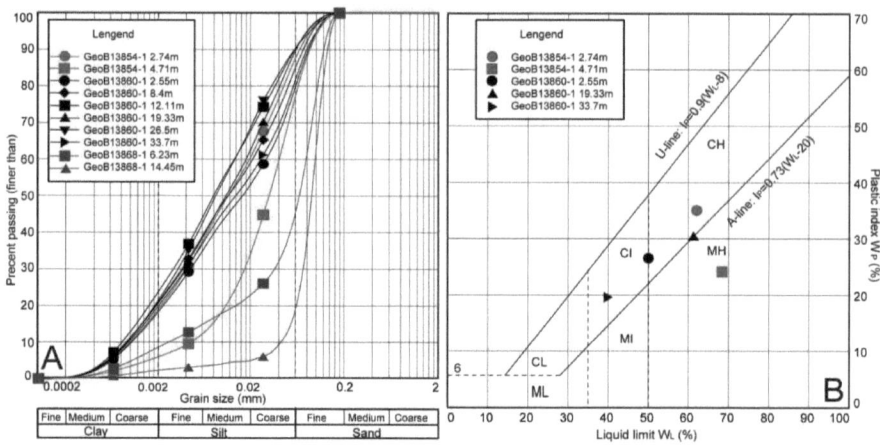

Fig. 2.4 *(A) Grain size distribution for sediments from NS (GeoB13854-1 and GeoB13860-1) and SC (GeoB13868-1). (B) Plasticity charts for classification of fine-grained sediments of NS (Classification chart modified after Craig 2004). The A-line separates the clays (C) from silts (M), the U-line represents the approximate upper limit of w_L and w_P combinations for natural soils, CH: inorganic clays of high plasticity, MH: inorganic silts of medium to high plasticity, CI: inorganic clays of medium plasticity, MI: inorganic silts of medium plasticity, CL: inorganic clays of low plasticity, ML: inorganic silts of low plasticity.*

Tab. 2.1 *Summary of laboratory oedometer test results. ρ_b: bulk density, σ'_{vh}: hydrostatic vertical effective stress, σ'_{pc}: preconsolidation stress, OCR: overconsolidation rate, C_c: compression index, c_v: coefficient of consolidation, k: coefficient of permeability.*

Area	Core	Depth (m)	ρ_b (g/cm³)	σ'_{vh} (kPa)	σ'_{pc} (kPa)	OCR	C_c	c_v (m²/s)	k (m/s)
NS	GeoB13854-1	2.74	1.57	15.32	195	12.73	0.389	8.8e-08	8.9e-10
		4.71	1.56	25.92	45	1.74	0.304	1.1e-08	3.6e-10
	GeoB13860-1	8.4	1.82	67.32	75	1.11	0.148	5.7e-08	5.2e-10
		12.11	1.86	102.29	20	0.20	0.143	2.2e-07	2.3e-09
		19.33	1.83	157.20	190	1.21	0.256	7.1e-08	5.0e-10
		26.5	1.84	217.85	32	0.15	0.125	5.2e-08	4.7e-10
		33.7	1.84	279.02	40	0.14	0.145	4.9e-08	2.8e-10
SC	GeoB13868-1	6.23	1.88	53.60	80	1.49	0.039	6.0e-08	2.6e-10
		14.45	1.96	135.52	18	0.13	0.014	2.9e-06	3.0e-09

2 Geotechnical characteristics and slope stability along the Uruguayan and northern Argentine margin

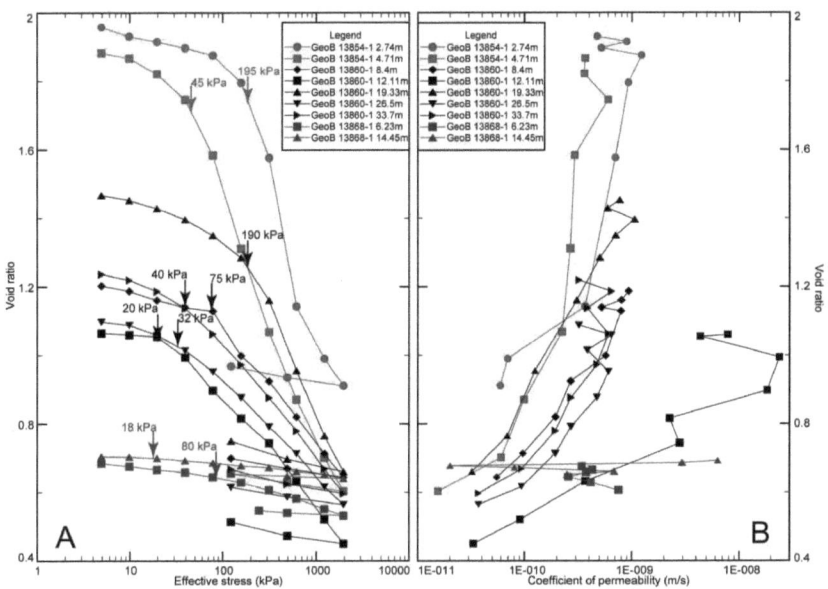

Fig. 2.5 *(A) e-log (σ'$_{vh}$) curves from oedometer tests with calculated σ'$_{pc}$ of NS (GeoB13854-1 and GeoB13860-1) and SC (GeoB13868-1). (B) e-k curves from oedometer tests.*

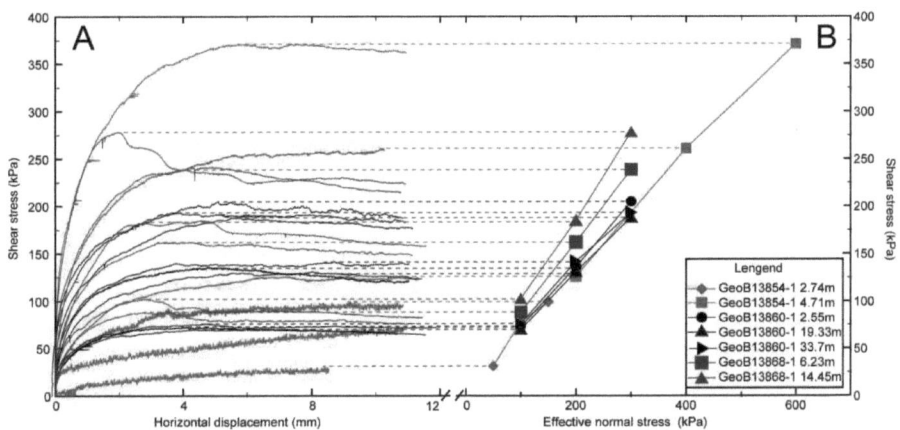

Fig. 2.6 *(A) Direct shear test protocols shown as shear stress versus horizontal displacement of NS (GeoB13854-1 and GeoB13860-1) and SC (GeoB13868-1). (B) Mohr-Coulomb failure planes of samples obtained from peak shear strength values.*

Tab. 2.2 Gibson parameters for overpressure estimation. c_v: coefficient of consolidation, m: sedimentation rate, t: duration of consolidation, T_g: Gibson time factor, $\lambda *$: overpressure ratio.

Area	Core	c_v (m²/s)	m (m/kyr)	t (kyr)	T_g	$\lambda *$
NS	GeoB13854-1	1.57e⁻⁰⁸	0.1	500	10.07	0.6
	GeoB13860-1	4.17e⁻⁰⁸	0.1	500	3.80	0.5
SC	GeoB13868-1	1.30e⁻⁰⁷	0.1	500	1.22	0.3

Tab. 2.3 Summary of direct shear test results. c': cohesion, ϕ': angle of internal friction.

Area	Core	Depth (m)	c' (kPa)	ϕ' (°)
NS	GeoB13854-1	0.74	1.52	34.25
		4.71	7.59	31.72
	GeoB13860-1	2.55	6.33	33.34
		19.33	11.53	30.55
		33.7	20.07	30.31
SC	GeoB13868-1	6.23	12.91	36.89
		14.45	12.92	41.30

2.4.2 Slope stability analysis

For the slope stability, both undrained and drained analyses were performed and ranges of factor of safety values were obtained. Factors of safety for four different scenarios were calculated using Equations 4-7 with two parameters changing within a certain range and other parameter keeping at constant value (Tab. 2.4; Fig. 2.7). According to the parameters defining the current situation, the factors of safety for undrained and drained static cases can be constrained. The slopes appear to be presently stable under both undrained and drained static conditions. Since slope geometry plays an important role in slope instability, variable slope angles (ranging from 1-5°) and variable failure depth levels (10-100 m) were used to calculate FS for the undrained static case. The undrained shear strength-depth relation was obtained using cone penetrometer data with linear regression. The results indicate that slope failure depth has less influence on slope stability than slope angle. For other scenarios with an assumed slope failure depth of 50 mbsf, slopes are stable in the undrained static case. SC is vulnerable with FS in the order of 1-2. For the drained static case, different overpressure ratios (0-0.9) and different slope angles (1-5°) were chosen for FS calculations. The results indicate overpressure is still unlikely to trigger slope failure in static scenarios. Factors of safety for NS tend to be lower compared to the undrained situation, while for SC it tends to be higher than that for the undrained scenario.

Pseudostatic infinite slope stability analysis represents a first order estimation of seismic ground accelerations that affect a given slope. The minimum horizontal acceleration coefficient required to trigger slope failure (FS = 1) was back-calculated based on Equations 6 and 7. For the undrained earthquake case, high values of horizontal acceleration ratio (k_e = 0.081-0.34) are needed to trigger slope failure for NS, while a low value of horizontal acceleration ratio (k_e = 0.34) is needed to trigger slope failure for SC. Given the scatter of results of overpressure estimation, a value of 0.5 is assumed in the drained earthquake case for both areas. The results for SC indicate a higher value of horizontal acceleration ratio ($k_e = 0.20$) is required in the drained case compared to the undrained case. For NS, two different trends are observed. For the shallow sediments (GeoB13854-1), the horizontal acceleration ratio gets slightly higher values ($k_e = 0.125$), while for the deeper sediments (GeoB13860-1), the horizontal acceleration ratio shows a substantial decrease (from 0.36 to 0.128).

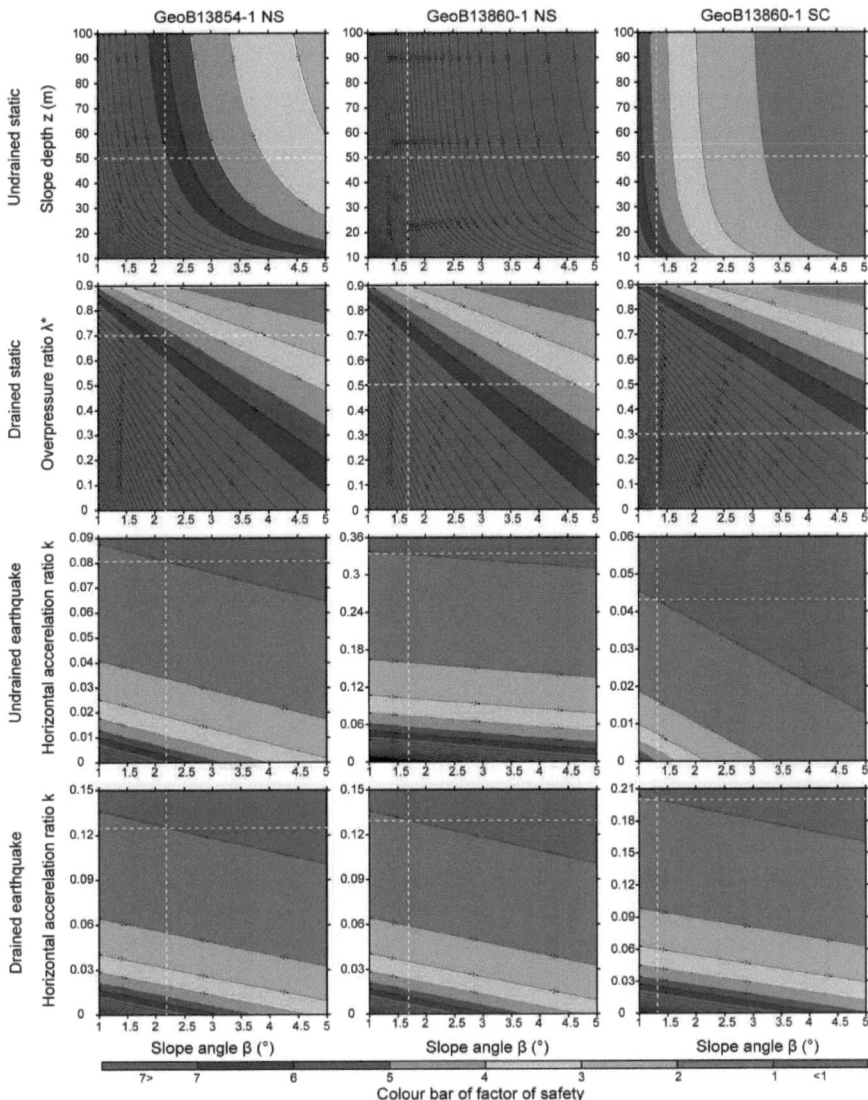

Fig. 2.7 Slope stability analyses and back-calculations of pseudostatic horizontal acceleration ratio showing FS as a function of slope angle and pseudostatic horizontal acceleration ratio under four

scenarios (undrained and drained, each static and earthquake conditions) for NS (GeoB13854-1 and GeoB13860-1) and SC (GeoB13868-1). Contour plots indicate FS values. Dashed white lines indicate current mean values of the parameters for static analyses and the values of pseudostatic horizontal acceleration required to trigger slope failure (FS = 1) at current mean value of the slope angle for pseudostatic analyses.

Tab. 2.4 Parameters used for slope stability calculations. US: undrained static, DS: drained static, UE: undrained earthquake, DE: drained earthquake; numbers in red colour indicate variable parameters.

Area	NS							SC			
Parameters	GeoB13854-1				GeoB13860-1				GeoB13868-1		
	US	DS	UE	DE	US	DS	UE	DE	US	DS	UE	DE
s_u (kPa)	0.99z+ 21.67		71.17		5.33z+ 45.57		312.07		0.92z+ 5.92		51.92	
z (m)	10-100		50		10-100		50		10-100		50	
β (°)		1-5				1-5				1-5		
c' (kPa)	-	4.5	-	4.5	-	12.5	-	12.5	-	13.0	-	13.0
φ' (°)	-	32.5	-	32.5	-	30.5	-	30.5	-	40.0	-	40.0
λ*	-	0-0.9	-	0.5	-	0-0.9	-	0.5	-	0-0.9	-	0.5
γ (kN/m^2)	-	-	15.21		-	-	18.05		-	-	19.23	
k_e	-	-	0-0.09	0-0.15	-	-	0-0.36	0-0.15	-	-	0-0.06	0-0.21
γ' (kN/m^2)		5.21				8.05				9.23		
g (m/s^2)		9.81				9.81				9.81		
FS/k_e	7.2/-	8.8/-	1/0.08	1/0.125	26/-	10/-	1/0.34	1/0.13	4.9/-	19.5/-	1/0.043	1/0.20

2.5 Discussion

2.5.1 Preconditioning factors

Preconditioning factors are defined as the physical and geotechnical properties of sediments resulting from initial deposition and post-depositional alteration, which promote slopes to be susceptible to instability. Physical and geotechnical properties depend to a large extent on lithology and grain size of the deposits. The grain size determines the value of pore volume-controlled properties (e.g., bulk density, water content, void ratio) (Baraza and Ercilla 1994). For undrained analysis, shear strength mainly depends on grain size and stress history of sediments. Slightly increasing sand contents could contribute a noted decrease in undrained shear strength (Lee et al. 1987). High overconsolidation ratio commonly results in higher undrained strength (Hampton et al. 1996). When translating those findings to our study, undrained analyses attest that SC is vulnerable because of its lower undrained shear strength due to coarse-grained sediments whereas NS is more stable because of higher undrained shear strength due to fine-grained sediments with overconsolidation (see Fig. 2.3).

For drained analysis, the shear strength is governed by cohesion (c'), angle of internal friction (φ') and effective vertical stress (σ'$_v$). Cohesion is regarded as a physico-chemical component of the shear strength, which is independent of the effective stress (Lamb and Whitman 1969). In general, fine-grained materials have higher cohesion comparing to coarse-grained materials. By contrast, the presence of fine-grained materials could be an important controlling parameter for a lower angle of internal friction of sediments (Huhn et al. 2006). The lowest ϕ' was measured for clayey sediments whereas sandy sediments show stronger frictional strength from the direct shear test. Effective vertical stress σ'_v is determined by pore pressure in the sediment. Some factors (e.g., high sedimentation rate, presence of bubble-phase gas, dissociation of gas hydrate, or fluid seepage) could result in overpressure (Sultan et al. 2004). Overpressure induced by sedimentary processes depends on permeability and sedimentation rate. Drained analysis show SC is more stable with higher angle of internal friction (~40°) and lower overpressure ratio (~0.3) whereas NS is more vulnerable with lower angle of internal friction (30.5-32.5°) and higher overpressure ratios (0.5-0.6). Sediments of NS have lower permeability than those of SC mainly due to grain size differences. It is assumed that sedimentation rate in the order of 0.1 m/kyr suffice to cause overpressure that could lower effective stresses by a magnitude that triggers slope stability.Our slope stability analysis results for both study areas suggest that the slopes appear to be stable under both undrained and drained static situation at present. Hence additional triggers are needed to cause slope failure.

2.5.2 Triggering mechanisms

Triggering mechanisms are termed the external stimuli that initiate the slope instability processes (Sultan et al. 2004). Compared to the preconditioning factors, they occur at shorter time scales. For instance, earthquakes are known to trigger large submarine mass movements as imposing the horizontal acceleration, which is usually confined to undrained analysis (Locat and Lee 2002). When considering acceleration induced earthquakes as a static parameter, it is reasonable to assume a drained pseudostatic model (Mulder et al. 1994). From our study sites, SC is more vulnerable in an undrained scenario where a 0.066 g value of effective earthquake peak ground acceleration (PGA = 0.043 g/0.65) is sufficient to trigger the slope failure. For NS, PGA in the order of 0.12-0.19 (k_e = 0.08-0.125) is required to trigger the slope failure under undrained as well as drained conditions. In order to explore plausible scenarios for such earthquake events that might induce seismic shaking in this intensity range along the Argentine-Uruguay margin, PGA is estimated using an empirical attenuation equation after Campbell and Bozorgnia (2008) that depend on the combination of magnitude and source distance of earthquake. In general, the study area shows low seismic activity (Fig. 2.1A). Over the past 160 y, only one earthquake with magnitude 5.2 is reported in 1988 in the NS area, having distances of 20 km to GeoB13860-1 and 80 km to GeoB13854-1. PGA induced by the 1988 earthquake (0.06 g) is still too low to trigger instability of GeoB13860-1 though it is nearest to the epicenter of 1988 earthquake (20 km) (Fig. 2.8). The attenuation relationship indicates that moderate M 4 near-field events in epicenter distances < 6 km or strong earthquakes (e.g., M = 7) in the far-field in epicentral distances < 45 km are required to trigger slope failure for SC. In the NS area, moderate earthquakes with M 4 adjacent to NS or strong earthquakes (e.g., M = 7) with epicentral distances < 15 km are required to trigger slope failure there.

Oversteepening plays a secondary role in slope instability with respect to external triggering mechanisms (e.g., earthquake) (McAdoo et al. 2000). Since FS decreases with increasing slope angle, one possible mechanism for slope failure could be rapid sediment accumulation on the slope oversteepening the upper of slope, or bottom currents undercutting the toe of slope. With the combined action of high sedimentation rate up to 1 m/kyr during the Holocene found at the thalweg of the

Querendi Canyon (Voigt et al., 2013) and contour currents continuously reworking the base of the canyon flanks, SC is vulnerable with respect to oversteepening under undrained static situation (Figs. 2.1D, 2.2D).

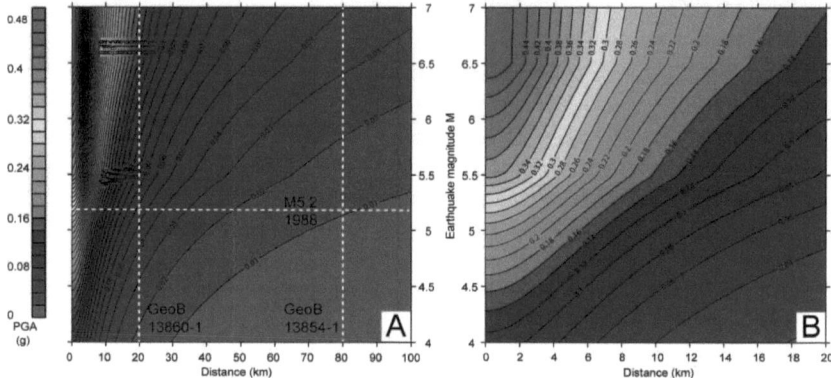

Fig. 2.8 (A) *PGA estimation using empirical attenuation equations after Campbell and Bozorgnia (2008) that depend on a combination of magnitude (M4-M7) and source distance (0-100 km) of the earthquake. Contour plots indicate the PGA. Dashed white line indicates the 1988 earthquake. Dashed red lines indicate the distances between the two slopes in NS (GeoB13854-1 and GeoB13860-1) to the epicenter the 1988 earthquake. (B) PGA estimation for source distance within 20 km.*

2.5.3 Slope failure modes between open slope versus canyon

From the above, the geotechnical analysis of two distinct areas on the continental slope of Uruguay and Argentina provides some valuable insights into the differences in slope failure mode between open slope and canyon.

Physical and geotechnical parameters of sediments as well as slope gradient are determined to a significant extent by grain size distribution. It is hence a product of the combined effects of sedimentological and oceanographic conditions that act on transport, sorting and accumulation of source materials (Adam et al. 1998). These processes then also control what precondition factors are met in a given depositional environment (slope, terrace and canyon sites; Fig. 2.1). Low gradient margins tend to have high sediment input and only a few modern canyons whereas high gradient margins tend to have lower sediment input but more modern canyons (O'Grady et al. 2000). Low gradient scenarios are likely promoting infrequent large-scale slope failures which support thick unstable sediment accumulation to build over long time in large area. In contrast, high gradient scenarios likely promote regular small-scale failure events which prevent thick unstable sediment accumulation to build through time (Migeon et al. 2011). Since canyons are important conduits for turbidity currents, the absence of major canyon systems on open slopes also promotes a depositional environment to form that allows rapid and widespread sediment accumulation (Locat and Lee 2009).

The continental slope between 1200-2800 m of NS is characterized by low slope gradient (1-3°), fine-grained sediments deposited at large areas, large-scale slope failure complex with high headwalls (up to 100 m) and acoustically transparent sediment bodies deposited at the foot of scarps. Northward bottom currents, which have carried the fine-grained sediments from the Rio de la Plata River, decrease in velocity in northward flowing direction so that clayey contourites are favored on the slope of NS. The geometry and structures of the NS slope failure complex is explained as retrogressive slide system with subsequent failures having progressively migrated farther upslope (Winkelmann et al. 2011). Slope stability analyses suggest that NS is currently stable and is unlikely to experience repeated small-scale failure but may reach unstable conditions if external triggers (e.g., earthquake) are strong enough to trigger large-scale failure. These results thus are consistent with general model of low gradient continental slope characterized by high sedimentation of fine-grained sediments and infrequent large-scale failures. In contrast, the continental slope between 1400-2500 m of SC is characterized by steep slopes (3-7°) and more abundant coarse-grained sediment on the flat Ewing terrace, which in the upper slope is dissected by the Querendi Canyon. In this area, here-in presented slope stability analyses attest the generally low stability, where steep slopes reflect repeated small-scale slope failures both during static conditions and certainly during infrequent seismic events. It is interpreted that high current velocities are responsible for deposition of sandy layers on the surface of upper slope of the canyon. Mass transport deposits are found in the mouth of the canyon but no distinct scars and mass transport deposits were found on the upper slope of the canyon (Krastel et al. 2011). This suggests that small slope failures continuously occur on the steep headwall, as also reveals in the slope stability calculation. Thereby the flanks of the canyon are constantly shaped by canyon wall incision and laterally retrograding into the open slope.

2.5.4 Slope failures along the Uruguayan and northern Argentine margin versus slope failures on other passive margins

An important result from this study is the findings that the investigated slopes along the Uruguayan and northern Argentine margin are mostly stable under static loading conditions and that slope failure can be initiated by additional driving forces during earthquake shaking. These findings suggest that earthquakes may play a critical role in initiating slope failure even on passive margin characterized by generally low seismicity levels. Similar conclusions have been reached from infinite slope stability and pseudostatic earthquake analysis on the slope instabilities of passive margins worldwide, such as upper slope of northeastern Australia (Puga-Bernabéu et al. 2013), continental slope off Uruguay (Henkel et al. 2011), northern Gulf of Mexico (Stigall and Dugan 2010), and Norwegian continental slope (Storegga slide; Leynaud et al. 2009; Leynaud et al. 2004). According to different precondition factors in different study areas (such as slope gradient, overpressure, grain size, weak layer etc.), critical peak ground acceleration generated by earthquake needed to initiate slope failure range between 0.02 g (Gulf of Mexico; Stigall and Dugan 2010), 0.1-0.2 g (Storegga slide; Kvalstad et al. 2005). These case studies further attest that overpressure generated by rapid sedimentation rates is important to precondition the slope towards low stability (e.g., Sultan et al. 2004), but in most cases overpressure alone would not have driven slope failure (Stigall and Dugan 2010; Kvalstad et al. 2005). Assuming overpressure ratios are 0.5-0.6 (estimated using Gibson's equation integrated consideration of sedimentation rate and permeability) on the upper 50 mbsf in the NS area of our study, PGAs exceeding 0.19-0.2 g is required to trigger sediment failure, suggesting comparable "triggering conditions" as for the Storegga landslide. In the Storegga region, weak layers have been identified in contouritic deposits that formed during interglacial periods and were rapidly buried under thick glacial marine deposits (Bryn et al. 2005). These fine-grained contouritic deposits are likely responsible for staircase appearance of headwalls and retrogressive behavior of the Storegga landslide. In our study area offshore Uruguay and northern Argentina, bottom currents are also responsible for the distinct morphologies between NS and SC. Currents also control sedimentation pattern, which in turn exert key control on slope stability precondition (see more detail in 2.5.3; Preu et al. 2013).

2.6 Conclusions

This paper presents sedimentological, physical and geotechnical results of the slope and canyon areas along the Uruguayan and Northern Argentine slope, which were used to compare the differences of preconditioning factors, triggering mechanisms and slope failure modes between open slope and canyon.

The NS region is characterized by a pattern of steep scarps on the gentle slope. With joint roles of fluvial discharge of the Rio de la Plata River and low-energy contour current, the sedimentary succession of NS is dominated by fine-grained material. Due to high undrained shear strength and moderate drained shear strength parameters, the slope along NS is stable and slope failure is unlikely under the current conditions. Slope failures are expected to occur for moderate earthquakes (M 4) in the NS are or strong events (e.g., M = 7) in epicentral distances < 15 km.

The SC region is characterized by slope failure at the canyon headwall and flanks with mass transport deposits stacked at the canyon mouth. Strong contour currents induced by the BMC are extensive reworking the downslope driven sediments and incorporating them into the sandy contourites. With low undrained shear strength, the slope of SC has a low stability under current conditions. Small-scale slope failures occur both during static conditions and certainly during infrequent seismic events such as moderate earthquakes in the near-field or strong earthquake (e.g., M = 7) in the far-field (epicentral distance < 45 km).

A comparison with submarine landslide studies on other passive continental margins reveals that different oceanographic and sedimentary settings result in different styles submarine mass movements in different passive continental margins (Krastel et al. accepted), while earthquake as additional triggering mechanism might be considerable for various settings. It is thus vital important that integration of geological, sedimentological, physical and geotechnical information to better understanding of submarine mass movements is approached by means of combined slope stability and earthquake analysis, as proposed in this study for the Uruguayan and northern Argentine margin.

Acknowledgements

We thank the captain and crew of the RV Meteor for their support during the cruise M78/3. Matthias Lange is thanked for outstanding technical assistance with the geotechnical laboratory devices. This study is funded through DFG-Research Center/Cluster of Excellence "The Ocean in the Earth System" as well as the Chinese Scholarship Council.

References

Adams EW, Schlager W, Wattel E (1998) Submarine slopes with an exponential curvature. Sedimentary Geology 117 (3–4): 135-141.

Assumpção M (1998) Seismicity and stresses in the Brazilian passive margin. Bulletin of the Seismological Society of America 88 (1): 160-169.

ASTM (2004a) Standard test method for direct Shear test of soils under consolidated drained conditions. ASTM International, Pennsylvania, United States.

ASTM (2004b) Standard test methods for one-dimensional consolidation properties of soils using incremental loading ASTM International, West Conshohocken, United States.

Baraza J, Ercilla G (1994) Geotechnical properties of near-surface sediments from the Northwestern Alboran Sea slope (SW mediterranean): Influence of texture and sedimentary processes. Marine Georesources & Geotechnology 12 (2): 181-200.

Blum P (1997) Physical Properties Handbook: A guide to the shopboard measurment of physical porperties of dee-sea cores. ODP Technical Notes 26: 118 pp.

Boyce RRE (1977) Deep Sea Drilling Project procedures for shear strength measurement of clayey sediment using modified Wykeham Farrance laboratory vane apparatus. Initial Reports DSDP, vol 36. U.S. Government Printing Office, Washington.

Bozzano G, Violante R, Cerredo M (2011) Middle slope contourite deposits and associated sedimentary facies off NE Argentina. Geo-Marine Letters 31 (5-6): 495-507.

Campbell KW, Bozorgnia Y (2008) NGA ground motion model for the geometric mean horizontal component of PGA, PGV, PGD and 5% damped linear elastic response spectra for periods ranging from 0.01 to 10 s. Earthquake Spectra 24: 139.

Casagrande A (1932) Research on the Atterberg limits of soils. Public roads 13 (8):121-136.

Casagrande A The determination of the pre-consolidation load and its practical significance. In: Proceedings of the 1st International Conference of Soil Mechanics and Foundation Engineering, 1936. pp 60-64.

Craig RF (2004) Craig's soil mechanics. Taylor & Francis.

Dugan B, Flemings PB (2002) Fluid flow and stability of the US continental slope offshore New Jersey from the Pleistocene to the present. Geofluids 2 (2): 137-146.

Flemings PB, Long H, Dugan B, Germaine J, John CM, Behrmann JH, Sawyer D, Scientists IE (2008) Pore pressure penetrometers document high overpressure near the seafloor where multiple submarine landslides have occurred on the continental slope, offshore Louisiana, Gulf of Mexico. Earth and Planetary Science Letters 269 (3-4): 309-325.

Franke D, Neben S, Ladage S, Schreckenberger B, Hinz K (2007) Margin segmentation and volcano-tectonic architecture along the volcanic margin off Argentina/Uruguay, South Atlantic. Marine Geology 244 (1-4): 46-67.

Garming JFL, Bleil U, Riedinger N (2005) Alteration of magnetic mineralogy at the sulfate-methane transition: Analysis of sediments from the Argentine continental slope. Physics of the Earth and Planetary Interiors 151 (3-4): 290-308.

Giberto DA, Bremec CS, Acha EM, Mianzan H (2004) Large-scale spatial patterns of benthic assemblages in the SW Atlantic: the Río de la Plata estuary and adjacent shelf waters. Estuarine, Coastal and Shelf Science 61 (1): 1-13.

Gibson R (1958) The progress of consolidation in a clay layer increasing in thickness with time. Geotechnique 8 (4): 171-182.

Hampton MA, Lee HJ, Locat J (1996) Submarine landslides. Reviews of Geophysics 34 (1): 33-59.

Harders R, Kutterolf S, Hensen C, Moerz T, Brueckmann W (2010) Tephra layers: A controlling factor on submarine translational sliding? Geochem Geophys Geosyst 11 (5): Q05S23.

Henkel S, Strasser M, Schwenk T, Hanebuth TJJ, Hüsener J, Arnold GL, Winkelmann D, Formolo M, Tomasini J, Krastel S, Kasten S (2011) An interdisciplinary investigation of a recent submarine mass transport deposit at the continental margin off Uruguay. Geochem Geophys Geosyst 12 (8): Q08009.

Hinz K, Neben S, Schreckenberger B, Roeser HA, Block M, Souza KGd, Meyer H (1999) The Argentine continental margin north of 48°S: sedimentary successions, volcanic activity during breakup. Marine and Petroleum Geology 16 (1): 1-25.

Huhn K, Kock I, Kopf A (2006) Comparative numerical and analogue shear box experiments and their implications for the mechanics along the failure plane of landslides. Norwegian Journal of Geology 86: 209-220.

Huppertz TJ (2011) Styles of continental margin sedimentation: comparing glaciated and non-glaciated slope systems using case studies on the southeast Canadian and northern Argentine and Uruguay continental slope. University of Bremen, Bremen.

Klaus A, Ledbetter MT (1988) Deep-sea sedimentary processes in the Argentine Basin revealed by high-resolution seismic records (3.5 kHz echograms). Deep Sea Research Part A Oceanographic Research Papers 35 (6):899-917.

Krastel S, Lehr J, Winkelmann D, Schwenk T, Preu B, Strasser M, Wynn RB, Georgiopoulou A, Hanebuth T Mass wasting along Atlantic continental margins: a comparison between NW-Africa and the de la Plata River region (northern Argentina and Uruguay). In: Krastel et al. (ed) Submarine Mass Movements and Their Consequences Advances in Natural and Technological Hazard Research, Kiel, accepted.

Krastel S, Wefer G, Hanebuth TJ, Antobreh A, Freudenthal T, Preu B, Schwenk T, Strasser M, Violante R, Winkelmann D (2011) Sediment dynamics and geohazards off Uruguay and the de la Plata River region (northern Argentina and Uruguay). Geo-Marine Letters 31 (4): 271-283.

Lamb TW, Whitman RV (1969) Soil mechanics. Massachusetts Institute of Technology.

Lee HJ, Chough SK, Jeong KS, Han SJ (1987) Geotechnical properties of sediment cores from the southeastern yellow sea: Effects of depositional processes. Marine Geotechnology 7 (1): 37-52.

Leynaud D, Mienert J, Nadim F (2004) Slope stability assessment of the Helland Hansen area offshore the mid-Norwegian margin. Marine Geology 213 (1-4): 457-480.

Leynaud D, Mienert J, Vanneste M (2009) Submarine mass movements on glaciated and non-glaciated European continental margins: A review of triggering mechanisms and preconditions to failure. Marine and Petroleum Geology 26 (5): 618-632.

Locat J, Lee H (2009) Submarine Mass Movements and Their Consequences: An Overview. In: Sassa K, Canuti P (eds) Landslides – Disaster Risk Reduction. Springer-Verlag Berlin Heidelberg, pp 115-142.

Locat J, Lee HJ (2002) Submarine landslides: advances and challenges. Canadian Geotechnical Journal 39 (1): 193-212.

Lonardi AG, Ewing M (1971) Sediment transport and distribution in the Argentine Basin. 4. Bathymetry of the continental margin, Argentine Basin and other related provinces. Canyons and sources of sediments. Physics and Chemistry of The Earth 8: 79-121.

Løseth TM (1999) Submarine massflow sedimentation: computer modelling and basin-fill stratigraphy. Springer.

Masson DG, Harbitz CB, Wynn RB, Pedersen G, Løvholt F (2006) Submarine landslides: processes, triggers and hazard prediction. Philosophical Transactions of the Royal Society A: Mathematical, Physical and Engineering Sciences 364 (1845): 2009-2039.

McAdoo BG, Pratson LF, Orange DL (2000) Submarine landslide geomorphology, US continental slope. Marine Geology 169 (1-2): 103-136.

Migeon S, Cattaneo A, Hassoun V, Larroque C, Corradi N, Fanucci F, Dano A, Mercier de Lepinay B, Sage F, Gorini C (2011) Morphology, distribution and origin of recent submarine landslides of the Ligurian Margin (North-western Mediterranean): some insights into geohazard assessment. Marine Geophysical Research 32 (1-2): 225-243.

Morgenstern N (1967) Submarine slumping and the initiation of turbidity currents. Marine geotechnique: 189-220.

Mulder T, Tisot J-P, Cochonat P, Bourillet J-F (1994) Regional assessment of mass failure events in the Baie des Anges, Mediterranean Sea. Marine Geology 122 (1-2): 29-45.

O'Grady DB, Syvitski JPM, Pratson LF, Sarg JF (2000) Categorizing the morphologic variability of siliciclastic passive continental margins. Geology 28 (3): 207-210.

Piola AR, Matano RP (2001) Brazil and Falklands (malvinas) Currents. In: Editor-in-Chief: John HS (ed) Encyclopedia of Ocean Sciences. Academic Press, Oxford, pp 340-349.

Piola AR, Matano RP, Palma ED, Möller OO, Jr., Campos EJD (2005) The influence of the Plata River discharge on the western South Atlantic shelf. Geophysical Research Letters 32 (1): L01603.

Preu B, Hernández-Molina FJ, Violante R, Piola AR, Paterlini CM, Schwenk T, Voigt I, Krastel S, Spiess V (2013) Morphosedimentary and hydrographic features of the northern Argentine margin: The interplay between erosive, depositional and gravitational processes and its conceptual implications. Deep Sea Research Part I: Oceanographic Research Papers 75 (0): 157-174.

Preu B, Schwenk T, Hernández-Molina FJ, Violante R, Paterlini M, Krastel S, Tomasini J, Spieß V (2012) Sedimentary growth pattern on the northern Argentine slope: The impact of North Atlantic Deep Water on southern hemisphere slope architecture. Marine Geology 329–331 (0): 113-125.

Puga-Bernabéu Á, Webster J, Beaman R (2013) Potential collapse of the upper slope and tsunami generation on the Great Barrier Reef margin, north-eastern Australia. Natural Hazards 66 (2): 557-575.

Seed HB (1979) Considerations in the earthquake-resistance design of earth and rockfill dams. Geotechnique 29 (3): 215-263.

Seed HB, Idriss IM (1971) Simplified procedure for evaluation soil liquefaction potential Soil Mechanics and Foundation Engineering SM 9: 1249-1273.

Sosa AB (1998) Sismicidad y sismotectónica en Uruguay. Física de la tierra (10): 167-186.

Stigall J, Dugan B (2010) Overpressure and earthquake initiated slope failure in the Ursa region, northern Gulf of Mexico. J Geophys Res 115 (B4): B04101.

Strasser M, Hilbe M, Anselmetti F (2011) Mapping basin-wide subaquatic slope failure susceptibility as a tool to assess regional seismic and tsunami hazards. Marine Geophysical Research 32 (1-2): 331-347.

Sultan N, Cochonat P, Canals M, Cattaneo A, Dennielou B, Haflidason H, Laberg JS, Long D, Mienert J, Trincardi F, Urgeles R, Vorren TO, Wilson C (2004) Triggering mechanisms of slope instability processes and sediment failures on continental margins: a geotechnical approach. Marine Geology 213 (1-4): 291-321.

ten Brink US, Barkan R, Andrews BD, Chaytor JD (2009) Size distributions and failure initiation of submarine and subaerial landslides. Earth and Planetary Science Letters 287 (1-2): 31-42.

Terzaghi K, Peck RB, Mesri G (1996) Soil mechanics in engineering practice. Wiley.

Voigt I (2013) Ocean circulation variability in the western South Atlantic during the Holocene. University of Bremen, Bremen.

Winkelmann D, Schwab J, Strasser M, Preu B, Schwenk T, Krastel S (2011) Large-Scale Submarine Mass-Wasting offshore Uruguay. EGU General Assembly 2011 Vol. 13 (Geophysical Research Abstracts):EGU2011-4603-2011.

Wood DM (1985) Some fall-cone tests. Géotechnique 38:64-68.

3 Submarine slope stability assessment of the central Mediterranean continental margin: the Gela Basin

F. Ai[1]*, J. Kuhlmann[1], K. Huhn[1], M. Strasser[2] and A. Kopf[1]

[1]MARUM-Center for Marine Environmental Sciences, and Faculty of Geosciences University of Bremen, Leobener Straße, 28359 Bremen, Germany.
[2]Geological Institute, ETH Zurich, Sonneggstrasse 5, 8092 Zurich, Switzerland.
*Corresponding author: aifei@uni-bremen.de

(Accepted by 6th International Symposium on Submarine Mass Movements and Their Consequences)

Abstract

This study investigates slope stability for a relatively small scale (5.7 km^2, 0.6 km^3), 8 kyr old landslide named Northern Twin Slide (NTS) at the slope of the Gela Basin in the Sicily Channel (central Mediterranean). The NTS is characterized by two prominent failure scars, forming two morphological steps of 110 m and 70 m height. Geotechnical data from a drill core upslope the failure scar (GeoB14403) recovered sediments down to ~53 m below seafloor (mbsf). The deposits show fine grain size (high clay content), high water content, low undrained shear strength and low internal friction angle, all of which suggests a weak layer around 28-45 mbsf that may act as potential slip surface in a future failure. Oedometer tests attest the sediments are highly underconsolidated and the average overpressure ratio λ^* is ~0.7. Slope stability analyses carried out for different scenarios indicate that the slope is stable both under static undrained and drained conditions. A relatively small horizontal acceleration of 0.03-0.08 g induced by an earthquake may be sufficient to cause failure. We propose that moderate seismic triggers may have been responsible for the twin slide formation and could also cause mass wasting in the future.

Keywords Slope stability, Submarine landslide, Geotechnical characteristics, Central Mediterranean

3.1 Introduction

Submarine landslides are ubiquitous on the continental margins. Overpressure near the seafloor plays a significant role in the occurrence of submarine landslides (Flemings et al. 2008). Seismic activity is considered as an important triggering mechanism of submarine landslides (Masson et al. 2006). Evaluating the effects of overpressure as a preconditioning factor and seismic shaking as trigger for submarine slope instability in earthquake-prone regions is compulsory for understanding the mechanisms of landslide initiation.

The study area is located between 200 and 800 m water depth along the eastern margin of the Gela Basin, of the Sicily Channel, central Mediterranean. There, the continental slope has a general gradient (appx. 3°) and steep slide head scarps (up to 32° in places) (Fig. 3.1A). Two recent slides (termed Northern twin slide and Southern twin slide) show subrounded scars on the upper slope and are characterized by bathymetric bulges at the base of the slope (Minisini et al. 2007). These landslides are described as multiple failures likely controlled by specific stratigraphic surfaces acting as glide planes (Minisini and Trincardi 2009; Trincardi and Argnani 1990).

In order to provide a possibility to assess future slope instabilities due to overpressure and seismic shaking, undrained and drained 1D infinite slope stability models are here introduced to model slope stability under both static and pseudostatic conditions.

3.2 Geological setting

The Gela Basin is the most recent (Plio-Quaternary) foredeep of the Maghrebian fold-and-thrust belt (Argnani et al. 1986). The extensional basin originated in the late Miocene to early Pliocene with the emplacement of the Gela nappe, which lasted until the early Pleistocene (Grasso 1993). Sequence stratigraphic interpretation on the shelf and upper slope area on the Gela Basin (Minisini and Trincardi 2009; Kuhlmann et al. this volume) identify, from top to bottom (Fig. 3.1B): (I) deposits resting on top of erosive unconformity ES1, (II) a progradational wedge pinching out towards NE and (III) deposits beneath sequence boundary SB1. In general, the studied northern slope of the Gela Basin shows high sedimentation rates are due to the interplay between abundant supply of fine-grained material and rapid subsidence (Emeis et al. 1996).

The NTS is characterized by two prominent failure scars forming two morphological steps of 110 m and 70 m height (Fig. 1B). The slope angle of the slide headwall is around 16°, while the surfaces of the displaced masses dip at 1.5-4.5° (Minisini et al. 2007). The accumulation area of the NTS is characterized by a morphologic bulge at the seafloor which extends 7 km downslope and 1.5 km in width. The source area of the NTS is 5.7 km^2 and the average height of the failure section is appx. 100 m. The runout of the NTS is 11.7 km (Minisini et al. 2007).

3.3 Material and Methods

3.3.1 Shipboard and laboratory analysis

The principal data set for this study is based on a MeBo (MARUM seafloor drill rig) core and a co-located gravity short core, both acquired during Cruise MSM15/3 in 2010. A 53 m-long succession from the undisturbed slope apron upslope the NTS scar was recovered. Visual core description was carried out on board shortly after core recovery on the split core. Discrete samples were taken on board to measure water content, density, porosity and void ratio using oven drying method and pycnometer (Blum 1997). Undrained shear strength (s_u) was estimated using a Mennerich Geotechnik (Germany) vane shear apparatus and Wykeham Farrance cone penetrometer (Wood 1985).

Laboratory experiments consisted of grain size distribution analysis using the Beckman Coulter LS 13320 particle size analyzer and Atterberg limits using the Casagrande apparatus and rolling thread method (Casagrande 1932). Consolidation tests were performed using uniaxial incremental loading oedometer system (ASTM 2004a), with permeability being estimated from the consolidation test results (Hüpers and Kopf 2012). The drained sediment strength parameters (cohesion c' and internal friction angle θ') were determined using drained direct shear tests (ASTM 2004b).

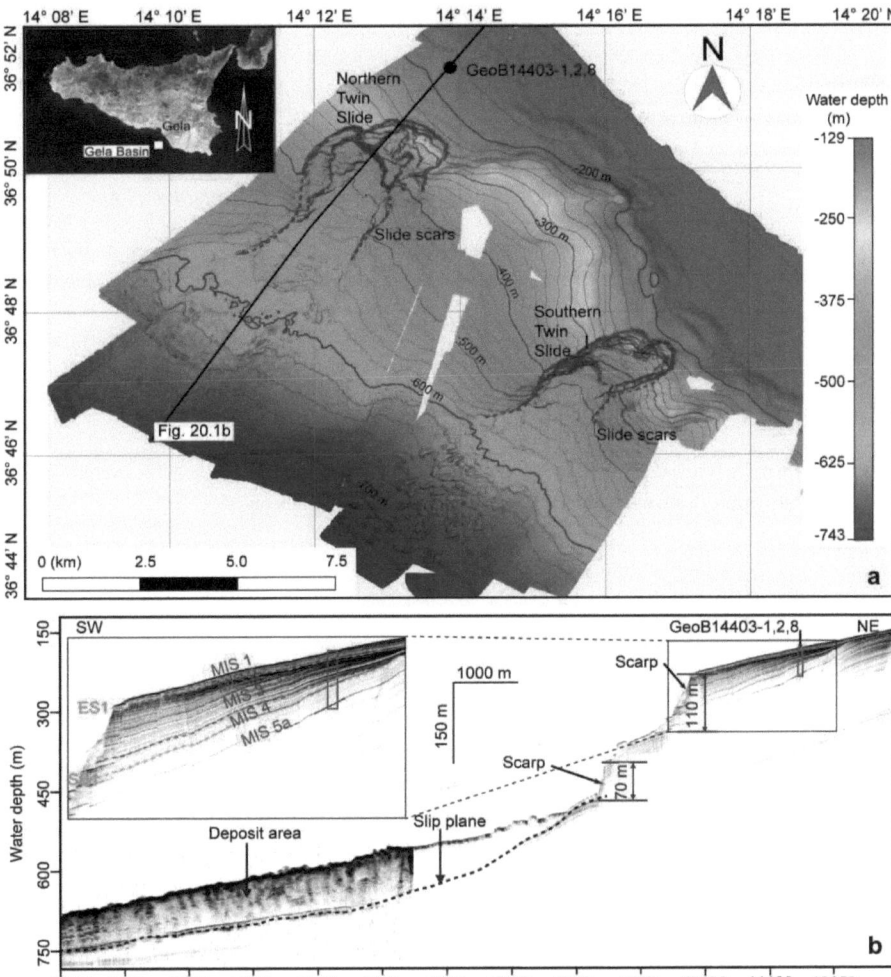

Fig. 3.1 *(A) Overview map showing the morphology of the Twin Slides in the Gela Basin offshore Sicily (Italy). Red dot indicates the location of core GeoB14403. Red dashed lines indicate the scarps of the Twin Slides. Red lines indicate the locations of parasound profiles presented in Fig. 1 B. (B) Parasound sub-bottom profile crossing the NTS. Red rectangle shows the location of the core GeoB14403. Green dashed lines indicate the Erosion surface 1 (ES1) and the sequence boundary 1 (SB1). Red dashed lines indicate the boundaries of different Marine Oxygen Isotope Stages (MIS) (Minisini et al. 2007; Kuhlmann et al. this volume). Black dashed line indicates the slip surface of the upper retrogressive landslide.*

3.3.2 Overpressure estimation

Overpressure (Δu) is defined as fluid pressure (u) in excess of hydrostatic equilibrium (u_0) (Dugan and Sheahan 2012). The role of overpressure is explained through Terzaghi's effective stress relationship:

$$\sigma_v' = \sigma_v - u = \sigma_v - (u_0 + \Delta u) = (\rho_b - \rho_w)gz - \Delta u = (\gamma - \gamma_w)z - \Delta u = \gamma'z - \Delta u \quad (1)$$

Where σ_v' is vertical effective stress, σ_v is total overburden stress, ρ_b is bulk density, ρ_w is density of water, γ is unit weight, γ_w is unit weight of water, γ' is buoyant weight, z is overburden depth, and g is the acceleration due to gravity.

Preconsolidation stress (σ_{pc}') interpreted from consolidation test is an approach to evaluate overpressure (Casagrande 1936):

$$\Delta u = \sigma_{vh}' - \sigma_{pc}' \quad (2)$$

Where σ_{vh}' is vertical effective stress for hydrostatic conditions ($\sigma_{vh}' = \sigma_v - \rho_w gz$). The overpressure was used to perform back analysis of slope stability under drained condition.

3.3.3 Slope stability analysis

The 1D infinite slope stability analysis is used to calculate the factor of safety (FS). In the infinite slope approximation FS ≥ 1 represents stability and FS ≤ 1 represents instability. For static conditions the FS calculation after Morgenstern (1967) and Løseth (1999) follows:

$$FS = \frac{S_u}{\gamma' z \sin\beta \cos\beta} \quad \text{(Undrained)} \quad (3)$$

$$FS = \frac{c' + r'z(\cos^2\beta - \lambda^*)\tan\theta'}{\gamma' z \sin\beta \cos\beta} \quad \text{(Drained)} \quad (4)$$

Where β is slope angle and λ* is overpressure ratio ($\lambda^* = \Delta u/\sigma_{vh}'$). The evaluation of slope stability under earthquake loading is commonly based on pseudostatic analysis (ten Brink et al. 2009; Morgenstern 1967). The overpressure that may be generated during an earthquake is not taken into account for the slope stability analysis. The seismic response included in the FS calculation is the integrated horizontal ground acceleration k g (where k is seismic coefficient and g is the acceleration due to gravity), which is assumed to be applied over a time period long enough for the induced shear stress can to be considered constant.

$$FS = \frac{S_u}{\gamma' z[\sin\beta \cos\beta + k(\gamma/\gamma')\cos^2\beta]} \quad \text{(Undrained)} \quad (5)$$

$$FS = \frac{c' + r'z(\cos^2\beta - \lambda^*)\tan\theta'}{\gamma' z[\sin\beta \cos\beta + k(\gamma/\gamma')\cos^2\beta]} \quad \text{(Drained)} \quad (6)$$

The aim of the stability back-analysis is to estimate the static FS of the slope and the minimum seismic acceleration required to trigger a slope failure under undrained and drained conditions, respectively.

3.4 Results

3.4.1 Physical and geotechnical properties

The physical properties and geotechnical results are presented in Fig. 3.2. The dominant lithology is homogeneous silty clay to clayey silt with a narrow range of particle sizes. The sediment's plastic limit is ~28%, while the liquid limit ranges from 55 to 74%. Natural water content is close to liquid limit and gradually decreases with depth, while bulk density gradually increases with depth. Undrained shear strength values range from a few kPa near the seafloor to ~50 kPa at depth, where the pocket penetrometer shows generally slightly higher values than vane shear. The undrained shear strength-depth relation obtained from linear regression is: $s_u = 0.6z+3.7$. The sediment appears to be underconsolidated as inferred from low ratios between the undrained shear strength and vertical effective stress at static condition ($S_u/\sigma_{vh}' \approx 0.1$) (Locat and Lee 2002).

Consolidation tests indicate sediments are slightly underconsolidated (OCR = $\sigma_{pc}'/\sigma_{vh}'$ = 0.76 at 4.08 mbsf) in shallow subsurface depth and trend to strongly underconsolidated at the deeper depth (OCR = 0.23 at 49 mbsf). Overpressures estimated from consolidation test results using Eq. 2 indicate that overpressure increases with depth and λ^* ranges from 0.24 to 0.77. Drained direct shear tests indicate that sediments from 28-45 mbsf have somewhat lower angle of internal friction (appx. 20°) compared to one sample from shallow depth (appx. 30°). Values of the effective cohesion intercept (c'), range from 0 to 30 kPa for granular to clay-rich sediments.

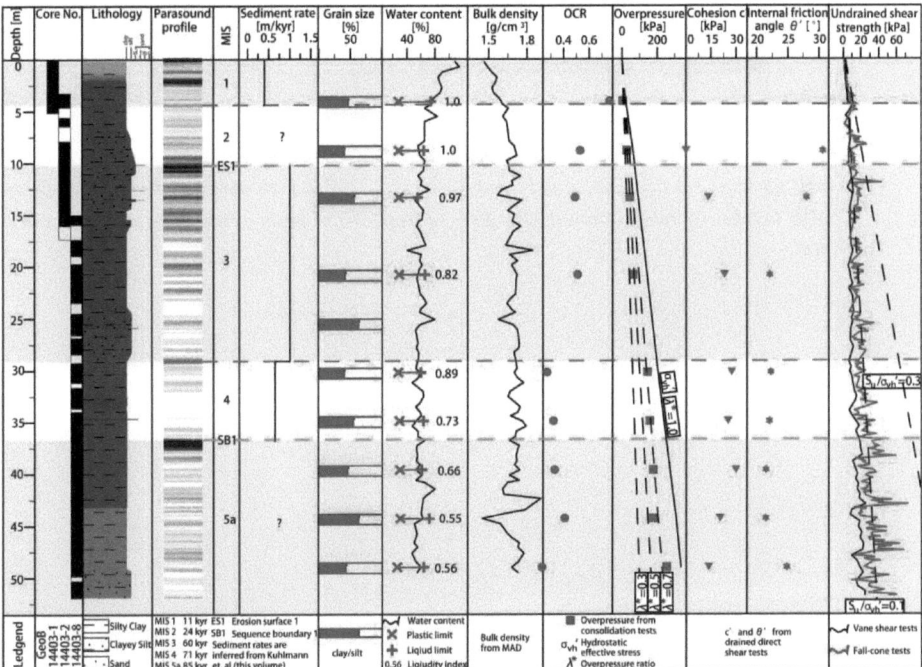

Fig. 3.2 Lithological, physical properties and geotechnical properties profile of core GeoB14403 for slope stability assessment.

3.4.2 Slope stability analysis

The back-analysis was carried out with two variables while the other parameters were kept at constant values (Tab. 3.1). Results of various scenarios of slope stability analysis have been tested (Tab. 3.1 and Figs. 3.3, 3.4). The slope appears to be presently stable both under static undrained and drained conditions. The depth of the failure surface shows less influence on FS compared to slope angle in the undrained slope stability analysis. Overpressure ratios of ≥ 0.93 are required to fail a drained slope at an angle of 5.5°. Pseudo-static analysis indicate that higher horizontal acceleration (k = 0.053) is required to trigger slope failure (FS = 1) at current mean values of the slope angle (β = 3°) in drained condition.

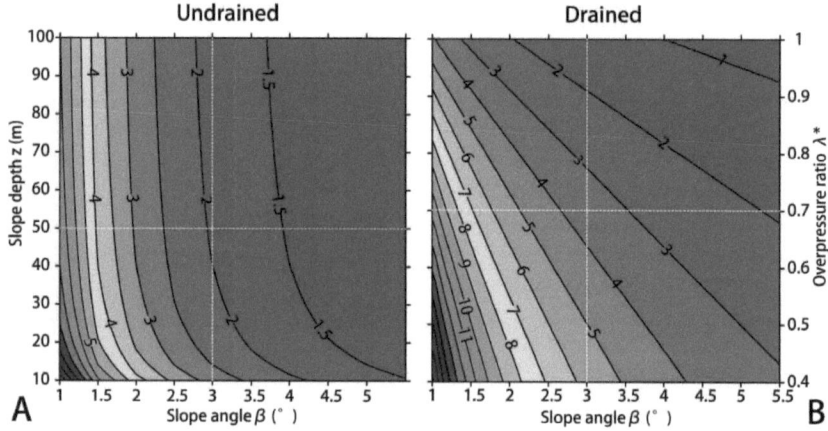

Fig. 3.3 Slope stability analysis under undrained (A) and drained (B) conditions for the NTS. Contour plots indicate FS values. Dashed white lines indicate current mean values of the parameters in the NTS.

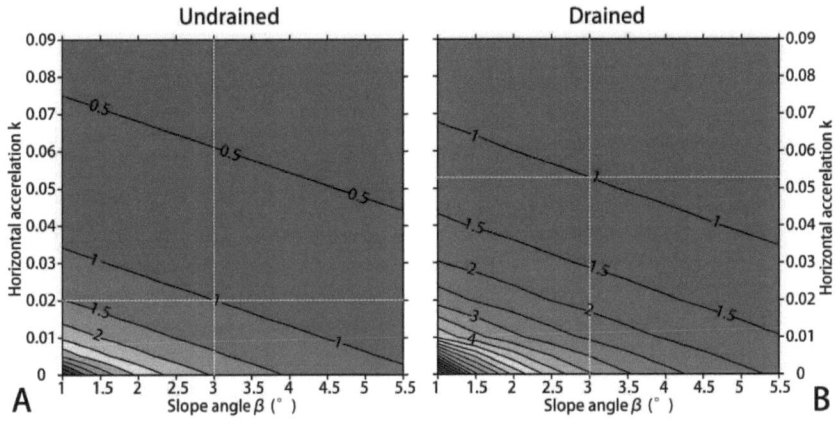

Fig. 3.4 *Back-analysis of slope stability showing FS as a function of slope angle and pseudostatic horizontal acceleration under undrained (A) and drained (B) conditions for the NTS. Contour plots indicate FS values. Dashed white lines indicate the value of pseudostatic horizontal acceleration required to trigger slope failure (FS = 1) at current mean value of the slope angle.*

Tab. 3.1 *Parameters used for slope stability calculations (data derived from Fig. 2)*

Input parameters	Static or variable/value (in different scenarios)			
	Undrained static	Drained static	Undrained seismic	Drained seismic
Undrained shear strength (kPa)	Static/0.6z+3.7		Static/33.7	
Depth of the failure surface, z (m)	Variable/10-100		Static/50	
Slope angle, β (°)	Variable/1-5.5			
Cohesion, c' (kPa)	-	Static/23.5	-	Static/23.5
Internal friction angle, θ' (°)	-	Static/21	-	Static/21
Overpressure ratio, λ*	-	Variable/0.4-1.0	-	Static/0.7
Unit weight, γ (kN/m^2)	-	-		Static/16.4
Horizontal acceleration, k	-	-	Variable/0-0.09	
Buoyant weight, γ' (kN/m^2)	6.6			
Gravitational acceleration, g (m/s^2)	9.81			
FS/k	1.9/0	3.6/0	1/0.02	1/0.053

3.5 Discussion

3.5.1 Preconditioning factors

Generally factors such as high sedimentation rate, slope steepening and the presence of intrinsically weak layer can greatly reduce the factor of safety of slope under static loading conditions. High sedimentation (≥ m/kyr) of low permeability sediments can generate overpressure, a condition known as underconsolidated (Dugan and Sheahan 2012). Undrained shear tests and consolidation tests presented here suggest the sediments are strongly underconsolidated. The state of overpressure in the Gela basin most likely results from deposition of fine-grained sediments with high sedimentation rates. Based on seismostratigraphic interpretation and shallow cores in the greater extremely high sedimentation rate (2.5-3.3 m/kyr) were proposed since the Last Glacial Maximum at 18-24 kyr B.P. (Minisini and Trincardi 2009). Kuhlmann et al. (this volume) estimates the sedimentation rate ranged ~0.5 m/kyr during MIS 1 and ~1 m/kyr during MIS 3 according to the MeBo core recovered from the undisturbed slope apron of the NTS. Based on our oedometer experiments, the coefficient of permeability of sediment is in the order of 10^{-10} m/s (permeability $\leq 10^{-16}$ m^2). Such a low permeability may inhibit fluid seepage and thereby induce the overpressure buildup (Dugan and Sheahan 2012).

The slope stability analysis suggests the slope angle has a larger influence on the FS compared to depth of failure plane in study area. Since FS decreases with increasing slope angle, one possible mechanism for failure could be rapid wedge-shaped sediment accumulation during sea level low stand (Minisini et al. 2007) or bottom currents leading to net erosion and undercutting the toe of slope (Bennett and Nelsen 1983). Both mechanisms have been hypothesis to occur in the study area (Minisini et al. 2007; Verdicchio and Trincardi 2008). Based on low internal friction angle derived from direct shear test, sediments ranged 28-45 mbsf appear to be weaker, potentially forming a preferential slip plane for future slope failure. Slope stability analysis further suggests the slope is stable in static conditions. To reach FS = 1 or lower, an overpressure ratio of 0.93 at slope angle of 5.5° would be required, which is not observed at present. Hence additional triggers are needed to generate slope failure.

3.5.2 Triggering factors

Pseudostatic analysis suggests that pseudostatic horizontal acceleration in the order of 0.02-0.053 g is required to fail the slope (FS = 1) in undrained as well as drained conditions. Strasser et al. (2011), based on the principles by Seed and Idriss (1971) and Seed (1979) suggested that the pseudostatic horizontal acceleration only represent ~65% of the effective earthquake peak ground acceleration (PGA = 0.03-0.08 g). In order to explore plausible scenarios for earthquake events, which might induce seismic shaking in this intensity range and thus could trigger retrogressive failure of the NTS, PGA is estimated using empirical attenuation equations after Bindi et al. (2011) that depend on combination of magnitude and source distance of earthquake (Fig. 3.5).

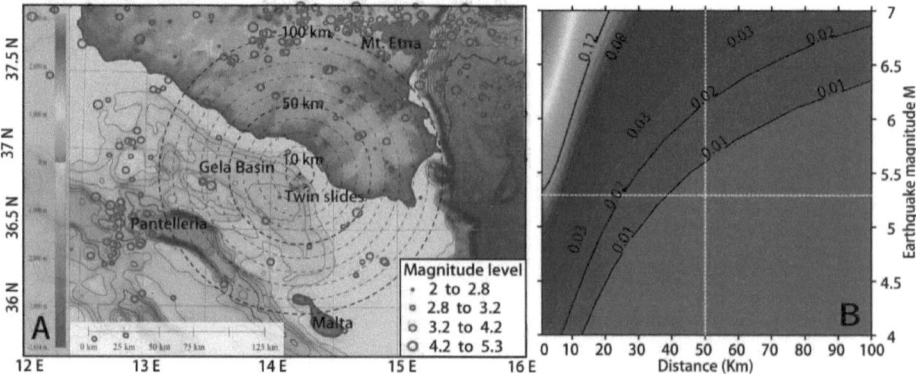

Fig. 3.5 (A) Earthquake record of the Sicily channel since 1970 (USGS database). Dashed black circles indicate the distance to the study area. Red circles indicate magnitude levels of earthquakes. (B) PGA estimation using empirical attenuation equations after Bindi et al. (2011) that depend on combination of magnitude and source distance of earthquake. Contour plots indicate the PGA. Red lines indicate the PGA values are required to trigger slope failure. Dashed white lines indicate the largest and closest earthquake in the study area during last 40 yr.

The Sicily channel shows less seismic activity (Fig. 3.5A). Over the last 40 yr, only small earthquakes occurred, which are too low to trigger instabilities (Fig. 3.5B). The attenuation relationship combined with the critical threshold condition for instability, as revealed from our analysis, indicates that moderate earthquakes with magnitudes of 4.0-4.8 near or at the location of the study area, or strong (M7), far-field events in epicentral distance < 20-80 km are required to trigger the slope failure. The presence of slope failure at the study area in the past indicates that over a longer time scale, larger magnitude earthquakes than those recorded during the instrumental period may have occurred. We propose that moderate seismic triggers may have been responsible for NTS formation and could also cause mass wasting in the future.

3.6 Conclusions

In summary, we have demonstrated geotechnical properties of sediments and slope stability analysis of the Gela Basin. Strongly underconsolidated of the sediments in the study area mainly attribute to rapid sedimentation in fine sediments. Despite of the high overpressure presented in the sediments, the stability analysis suggest the slope is stable under static conditions. Slope failure may be triggered by moderate earthquake (M4.0-M4.8) in the study area, or even strong events if farther away. Additional studies of in-situ pore pressure measurement and pore pressure change during the seismic loading are needed in order to further investigate the dynamic response and better assess the slope stability.

Acknowledgments

We thank the captain and crew of the RV Meteor for their support during the cruise MSM 15/3. Matthias Lange is thanked for outstanding technical assistance with the geotechnical laboratory devices. This study is funded through DFG-Research Center/Cluster of Excellence "The Ocean in the Earth System" as well as the Chinese Scholarship Council. We also like to acknowledge the reviewers, Brandon Dugan and Vasilios Lykousis, for their constructive remarks.

References

Argnani A, Cornini S, Torelli L, Zitellini N (1986) Neogene-Quaternary foredeep system in the Strait of Sicily. Mem Soc Geol It 36: 123-130.

ASTM (2004a) Standard test methods for one-dimensional consolidation properties of soils using incremental loading (Standard D2435-04). ASTM International, West Conshohocken, United States p. 10.

ASTM (2004b) Standard test method for direct shear test of soils under consolidated drained conditions (Standard D3080-04). ASTM International, West Conshohocken, United States p. 7.

Bennett RH, Nelsen TA (1983) Seafloor characteristics and dynamics affecting geotechnical properties at shelfbreaks. SEPM special publication 33: 333-355.

Bindi D, Pacor F, Luzi L, Puglia R, Massa M, Ameri G, Paolucci R (2011) Ground motion prediction equations derived from the Italian strong motion database. Bulletin of Earthquake Engineering 9 (6): 1899-1920.

Blum P (1997) Physical Properties Handbook: A guide to the shipboard measurement of physical properties of deep-sea cores. ODP Technical Notes 26: 118.

Casagrande A (1932) Research on the Atterberg Limits of soil. Public Roads 13(8): 121-136.

Casagrande A (1936) The determination of the pre-consolidation load and its practical significance. Proceedings of the 1st International Conference of Soil Mechanics and Foundation Engineering 3: 60-64.

Dugan B, Sheahan TC (2012) Offshore sediment overpressures of passive margins: Mechanisms, measurement, and models. Rev Geophys 50 (3): RG3001.

Emeis K, Robertson A, Richter C (1996) Site 963. Proceedings ODP, Initial Report 160: 55-84.

Flemings PB, Long H, Dugan B, Germaine J, John CM, Behrmann JH, Sawyer D, Scientists IE (2008) Pore pressure penetrometers document high overpressure near the seafloor where multiple submarine landslides have occurred on the continental slope, offshore Louisiana, Gulf of Mexico. Earth and Planetary Science Letters 269 (3-4): 309-325.

Grasso M (1993) Pleistocene structures along the Ionian side of the Hyblean Plateau (SE Sicily): implications for the tectonic evolution of the Malta Escarpment. In: Proc. Int. Sc. Meeting: 49-54.

Hüpers A, Kopf AJ (2012) Data report: consolidation properties of silty claystones and sandstones sampled seaward of the Nankai Trough subduction zone, IODP Sites C0011 and C0012. In: Proc. IODP Volume 322.

Løseth TM (1999) Submarine massflow sedimentation: computer modelling and basin-fill stratigraphy. Springer.

Locat J, Lee HJ (2002) Submarine landslides: advances and challenges. Canadian Geotechnical Journal 39 (1): 193-212.

J. Kuhlmann, A. Asioli, M. Strasser, F. Trincardi, K. Huhn (this volume) Integrated stratigraphic and morphological investigation of the Twin Slide complex offshore southern Sicily.

Masson DG, Harbitz CB, Wynn RB, Pedersen G, Løvholt F (2006) Submarine landslides: processes, triggers and hazard prediction. Philosophical Transactions of the Royal Society A: Mathematical, Physical and Engineering Sciences 364 (1845): 2009-2039.

Minisini D, Trincardi F (2009) Frequent failure of the continental slope: The Gela Basin (Sicily Channel). J Geophys Res 114 (F3): F03014.

Minisini D, Trincardi F, Asioli A, Canu M, Foglini F (2007) Morphologic variability of exposed mass-transport deposits on the eastern slope of Gela Basin (Sicily channel). Basin Research 19 (2): 217-240.

Morgenstern N (1967) Submarine slumping and the initiation of turbidity currents. Marine geotechnique: 189-220.

Seed HB (1979) Considerations in earthquake-resistant design of earth and rock-fill dams. Géotechnique 29: 215-263.

Seed HB, Idriss IM (1971) Simplified procedure for evaluating soil liquefaction potential. J Soil Mech

Found Div Proc Am Soc Civil Eng 97: 1249-1273.

Strasser M, Hilbe M, Anselmetti F (2011) Mapping basin-wide subaquatic slope failure susceptibility as a tool to assess regional seismic and tsunami hazards. Marine Geophysical Research 32: 331-347.

ten Brink US, Lee HJ, Geist EL, Twichell D (2009) Assessment of tsunami hazard to the U.S. East Coast using relationships between submarine landslides and earthquakes. Marine Geology 264 (1-2): 65-73.

Trincardi F, Argnani A (1990) Gela submarine slide: A major basin-wide event in the plio-quaternary foredeep of Sicily. Geo-Marine Letters 10 (1): 13-21.

Wood DM (1985) Some fall-cone tests. Géotechnique 38: 64-68.

Verdicchio G, Trincardi F (2008) Mediterranean shelf-edge muddy contourites: examples from the Gela and South Adriatic basins. Geo-Marine Letters 28 (3): 137-151.

4 Geotechnical characteristics and slope stability analysis on the deeper slope of the Ligurian margin, Southern France

Fei Ai[1ˣ]*, Annika Förster*[2]*, Sylvia Stegmann*[1] *and Achim Kopf*[1]

[1]MARUM-Center for Marine Environmental Sciences, and Faculty of Geosciences University of Bremen, Leobener Straße, 28359 Bremen, Germany

[2]Institute of Geosciences and Geography, Martin-Luther-University Halle-Wittenberg, von Seckendorff Platz 3, 06120 Halle (Saale), Germany

ˣCorresponding author: aifei@uni-bremen.de

(Submitted to the World Landslide Forum 3)

Abstract

Submarine slope failures of various types and sizes are common along the tectonic and seismically active Ligurian margin, northwestern Mediterranean Sea, primarily because of seismicity up to ~M6, rapid sediment deposition in the Var fluvial system, and steepness of the continental slope (average 11°). We present geophysical, sedimentological and geotechnical results of two distinct slides in water depth >1500 m: one located on the flank of the Upper Var Valley called Western Slide (WS), another located at the base of continental slope called Eastern Slide (ES). WS is a superficial slide characterized by a slope angle of ~4.6° and shallow scar (~30 m) whereas ES is a deep-seated slide with a lower slope angle (~3°) and deep scar (~100 m). Both areas mainly comprise clayey silt with intermediate plasticity, low water content (30-75 %) and under-consolidation to strong overconsolidation. Upslope undeformed sediments have low undrained shear strength (0-20 kPa) increasing gradually with depth, whereas an abrupt increase in strength up to 200 kPa occurs at a depth of ~3.6 m in the headwall of WS and ~1.0 m in the headwall of ES. These boundaries are interpreted as earlier failure planes that have been covered by hemipelagite or talus from upslope after landslide emplacement.

Infinite slope stability analyses indicate both sites are stable under static conditions; however, slope failure may occur in undrained earthquake condition. Peak earthquake acceleration from 0.09 g on WS and 0.12 g on ES, i.e. M5-5.3 earthquakes on the spot, would be required to induce slope instability. Different failure styles include rapid sedimentation on steep canyon flanks with undercutting causing superficial slides in the west and an earthquake on the adjacent Marcel fault to trigger a deep-seated slide in the east.

Keywords *submarine slope failure, geotechnical characteristics, slope stability analysis, Ligurian margin*

4.1 Introduction

Submarine slope failures represent the main agents of sediment transport from continental slope to deep ocean, and one of the most common geohazards impacting on both offshore infrastructures (e.g., pipeline, cables and platforms) and coastal areas (e.g., slope failure-induced tsunamis) (Locat and Lee 2002). Slope failures are generally controlled by long-term preconditioning factors (e.g., high sedimentation rate, weak layer and oversteepening) and short-term triggering mechanisms (e.g., earthquake, anthropogenic activity) (Sultan et al. 2004). However, the exact causes for the different slope failure styles are still poorly understood.

The Ligurian margin, northwestern Mediterreanean Sea, is one of most intensely studied natural laboratories for landslide initiation in seismically active areas because of its steep topography with numerous landslide scars of different size. Previous slope stability analyses in the region mainly focused on the 1979 Nice Airport Slide or the upper slope of Ligurian margin (Cochonat et al. 1993; Mulder et al. 1994; Sultan et al. 2004; Dan et al. 2007; Leynaud and Sultan 2010; Stegmann et al. 2011). This study presents two distinct slides (WS and ES) along the deeper slope of Ligurian margin (1500-2000 m below seaflow (mbsf)). Klaucke and Cochonat (1999) and Migeon et al. (2011) concentrated on the mophologies of slope failure and qualitatively identified their triggering mechanisms. Kopf et al. (2008) and Förster et al. (2010) charatererized the architecture and evolution of the slope failures. Our study presents geotechnical properties of sediments from undeformed, headwall and deposit areas of WS and ES. Those results are used for infinite slope stability of undeformed sediments under various conditions to (i) identify the preconditioning factors and (ii) quantify the influence of earthquakes as a key factor in slope failing mechanisms in this densely populated area.

4.2 Geological, geomorphological and lithological setting

The Ligurian Basin is considered as a back-arc basin that formed by continental rifting and drifting during the late Oligocene from the southeastward rollback of the Apennines-Maghrebides subduction zone (Larroque et al. 2012 and references therein). Currently, active basin deformation occurs offshore at a slow rate of ~1.1 m/ka NNW-SSE, which involves moderate seismic activity with common earthquake magnitudes of M2.2 to M4.5 (Fig. 4.1A). However, earthquake magnitudes up to M6.8 (e.g., 1887 Ligurian earthquake) are documented at the Ligurian margin (Larroque et al. 2012). The Marcel Fault shows evidence of present activity that three moderate earthquakes (M3.8-M4.6) took place around this fault over the last 30 years (Larroque et al. 2012 and Fig. 4.1B).

The northern upper continental slope of the Ligurian Basin is eroded by two major canyons (Var canyon and Paillon canyon), which coalesce at a depth of 1650 m (Cochonat et al. 1993 and Fig. 4.1B). A single channel was formed at the confluence of the two canyons and divided into three parts: upper, middle and lower valleys. The walls of Upper Valley are highly dissected by small retrogressive failure events (Migeon et al., 2011) such as that west of Cap Ferrat Ridge called Western Slide (WS). It is characterized by shallow headwalls (< 30 m) with high slope gradients of ~4.6° (Fig. 14.C and Fig. 4.2A). A slope failure east of Cap Ferrat Ridge is termed Eastern slide (ES) and shows deep slide scars (80-120 m) and a lower slope gradient of ~3° (Fig. 4.1C and Fig. 4.2B).

Recent processes of sediment transport and deposition in the Var Upper Valley were mainly dominated by hyperpycnal-flow activity, failure-induced turbidity currents, and hemipelagite emplacement (Migeon et al. 2011). The lithostratigraphic succession of WS is characterized by homogenous, fine-grained hemipelagic clayey silt with some coarse-grained sand intervals (Kopf et al. 2008 and Fig. 4.3A). Areas east of Cap Ferrat Ridge are not connected to major fluvial input of the Var system and receive only hemipelagic sediments (Klaucke et al. 2000). The sediments are generally composed of well-bioturbated, homogenous, fine-grained hemipelagic deposits (Kopf et al. 2008 and Fig. 4.3B).

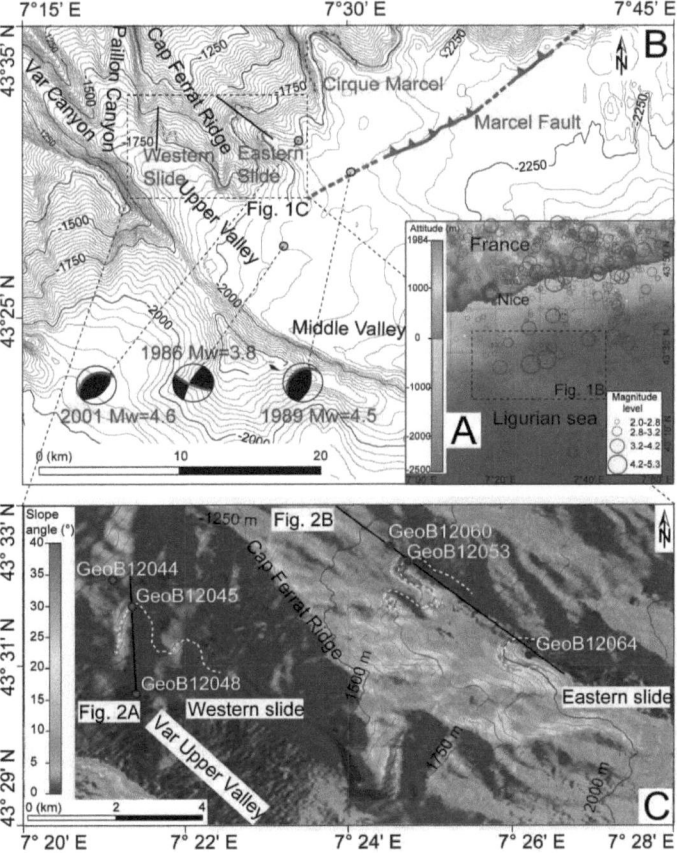

Fig. 4.1 *(A) Map showing the location of the study area, red circles indicate earthquake records of the Ligurian margin from 1980 to 2010 (catalogue from the Bureau Central Sismologique Français). (B) Bathymetric map of deeper slope of Ligurian margin with focal mechanisms of the moderate earthquakes associated with the Marcel Fault (taken from Larroque et al., 2012). (C) Gradient map of WS and ES. Circles indicate core locations. Dashed red lines mark the headwalls of both slides (revised after Förster et al. 2010). Black lines indicate the locations of seismic profiles shown in Fig. 4.2A and Fig. 4.2B.*

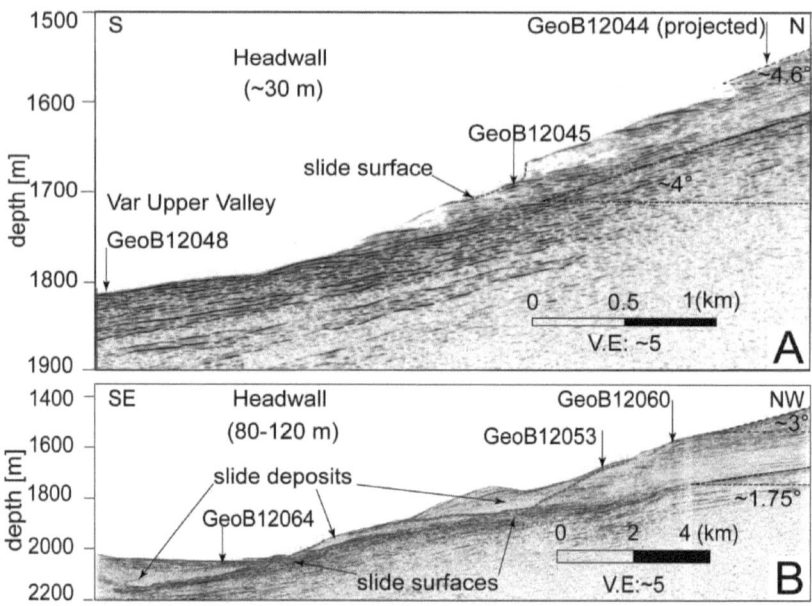

Fig. 4.2 *Seismic profiles of WS (A) and ES (B) (modified after Förster et al. 2010). Note that the bulge in panel B does not show the real morphology of the headwall but is an artefact because the profile crosses the flank of slope.*

4.3 Materials and Methods

4.3.1 Laboratory tests

The principal data set for this study is based on six gravity cores from undeformed slope, headwall and deposit areas of the WS and ES events. Water content was measured by GeoTeK Multi Sensor Core Logger (MSCL) on the archive halves at 2 cm intervals. Undrained shear strength (S_u) was estimated using a Mennerich Geotechnik (Germany) vane shear apparatus and Wykeham Farrance cone penetrometer. Grain size distribution analysis using the Beckman Coulter LS 13320 particle size analyzer and Atterberg limits using the Casagrande apparatus and rolling thread method were carried out. Oedometer tests were performed using a GIESA uniaxial incremental loading oedometer system. The drained sediment strength parameters (cohesion c' and internal friction angle ϕ') were determined using a displacement-controlled direct shear apparatus built by GIESA (Germany).

4.3.2 Slope stability analysis

The 1D infinite slope stability analysis is used to calculate the factor of safety (FS). For static conditions the FS calculation after Morgenstern (1967) follows:

$$FS = \frac{S_u}{\gamma' z \sin\theta \cos\theta} \quad \text{(undrained)} \qquad (1)$$

$$FS = \frac{c' + \gamma' z \left(\cos^2\theta - \lambda^*\right)\tan\phi'}{\gamma' z \sin\theta \cos\theta} \quad \text{(drained)} \qquad (2)$$

Where θ is slope angle and λ* is overpressure ratio ($\lambda^* = \Delta u/\sigma'_{vh}$), Δu is overpressure, σ'_{vh} is vertical effective stress for hydrostatic conditions ($\sigma'_{vh} = \gamma' z$). γ' is buoyant weight, z is overburden depth. Pseudostatic analysis was used for evaluation of slope stability under earthquake, which is assumed the integrated horizontal ground acceleration k g (where k is the seismic coefficient and g is the acceleration due to gravity) to be applied over a time period long enough for the induced shear stress to be considered constant while the overpressure that may be generated during an earthquake is not taken into account for the slope stability analysis (see Mulder et al. 1994):

$$FS = \frac{S_u}{\gamma' z \left[\sin\theta\cos\theta + k(\gamma/\gamma')\cos^2\theta\right]} \quad \text{(undrained)} \qquad (3)$$

$$FS = \frac{c' + \gamma' z \left(\cos^2\theta - \lambda^*\right)\tan\phi'}{\gamma' z \left[\sin\theta\cos\theta + k(\gamma/\gamma')\cos^2\theta\right]} \quad \text{(drained)} \qquad (4)$$

Where γ is unit weight.

4.4 Results

4.4.1 Physical and geotechnical properties

Water content and undrained shear strength of sediments are presented in Figure 4.3. Sediments from undeformed slopes have high values of water content (~60 %), while lower values (~30 %) are seen in deeper parts of sediment cores from the headwall. Sediments from the ES deposit area have similar water content as sediments from the undeformed upslope region. Low water content (~30 %) of sediments from the deposit area of WS is attributed to coarse-grained materials. Undrained shear strength of sediments from undeformed slope gradually increase with depth to ~20 kPa at 5 m core depth with value of S_u/σ'_{vh} ranging between 0.2-0.4 which indicate normal consolidated state for marine sediments. Sediments from headwall have low shear strength (0-20 kPa) and increase rapidly up to ~200 kPa at 3.65 m for WS and 1.0 m for ES.

The dominant lithology is clayey silt (with ~20% clay) with an intermediate plasticity according to our Atterberg limit measurements. Oedometer tests indicate sediment from the undeformed slope of WS is underconsolidated (overconsolidation ratio (OCR) = $\sigma'_{pc}/\sigma'_{vh}$) = 0.62, λ*= 1 - OCR = 0.38) and normally consolidated (OCR = 0.99, λ* = 0) for ES whereas sediments below the slip surface near the headwall are strongly overconsolidated (OCR = 9.6 for WS, OCR= 72.2 for ES) (Fig. 4.4). The calculated thickness of removed overburden material are 31 m for WS and 100 m for ES using the equation of Silva et al. (2001). This is consistent with the depth estimates based on seismic profiles (Fig. 4.2). Drained direct shear test results are presented in Figure 4.5. Values of c' are lower in sediments from WS (1.8-7.7 kPa) than in ES (6.7-10.7 kPa) whereas values of φ' are slightly higher in sediments from WS (30.9-33.8°) than in ES (29.5-31.9°).

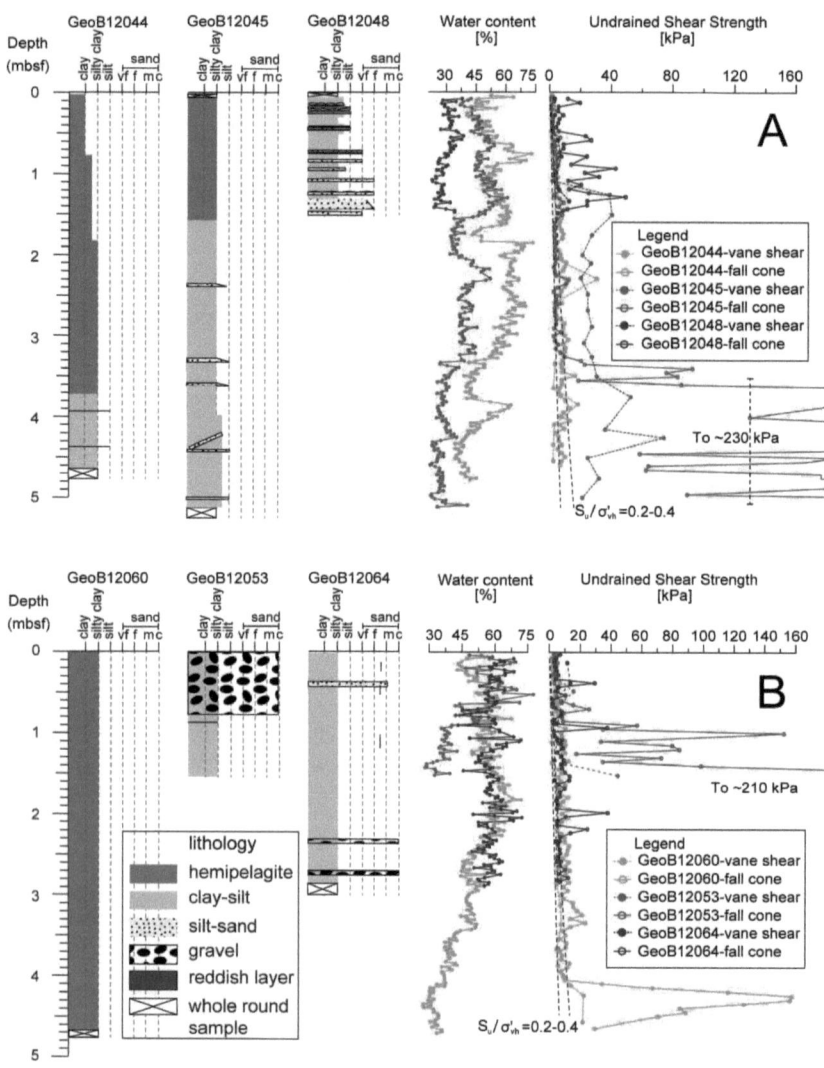

Fig. 4.3 Lithology, Water content as represntive for physical properties and undrained shear strength of the sediments from WS (A) and ES (B).

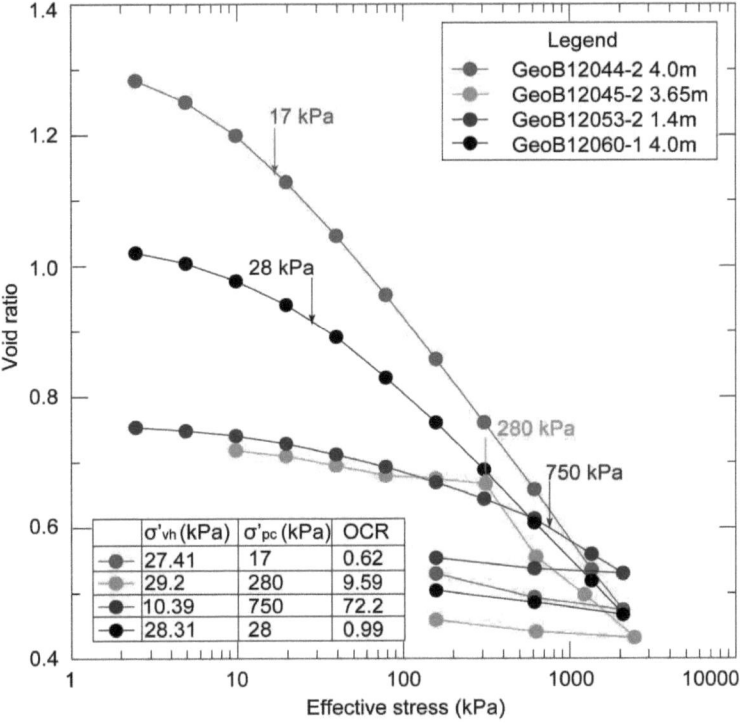

Fig. 4.4 *e-log (σ'v) curves from oedometer tests with calculated preconsolidation stress (σ'pc) and overconsolidation ratio (OCR).*

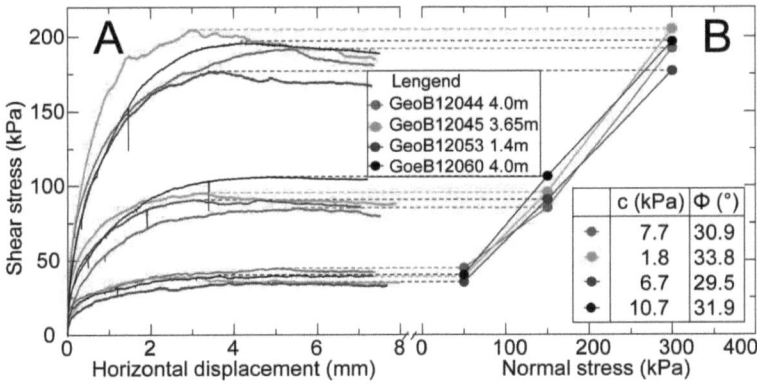

Fig. 4.5 *(A) Direct shear test protocols shown as shear stress versus horizontal displacement. (B) Mohr-Coulomb failure planes obtained from peak shear strength values.*

4.4.2 Slope stability analysis

Factors of safety for four different scenarios were calculated using Equations 1-4 with two parameters changing within a certain range while all others were kept constant (for details see Tab. 4.1). The undrained shear strength-depth relation was obtained using fall cone tests data with linear regression. We assume λ* of WS is 0.38 due to underconsolidated state and no overpressure in ES because of its normal consolidation state. Our data suggest that both slopes appear to be presently stable under both undrained and drained static conditions. The results further indicate that the slope angle has a stronger influence on slope stability than slope failure depth (Fig. 4.6). Pseudostatic infinite slope stability analysis represents a first-order estimation of seismic ground accelerations that affect a given slope. The minimum horizontal acceleration coefficient required to trigger slope failure (FS = 1) was back-calculated based on Equations 3 and 4. For the undrained earthquake case, a value of k = 0.08 is needed to trigger slope failure for ES, while a lower value of k = 0.06 is needed to fail the WS slope.

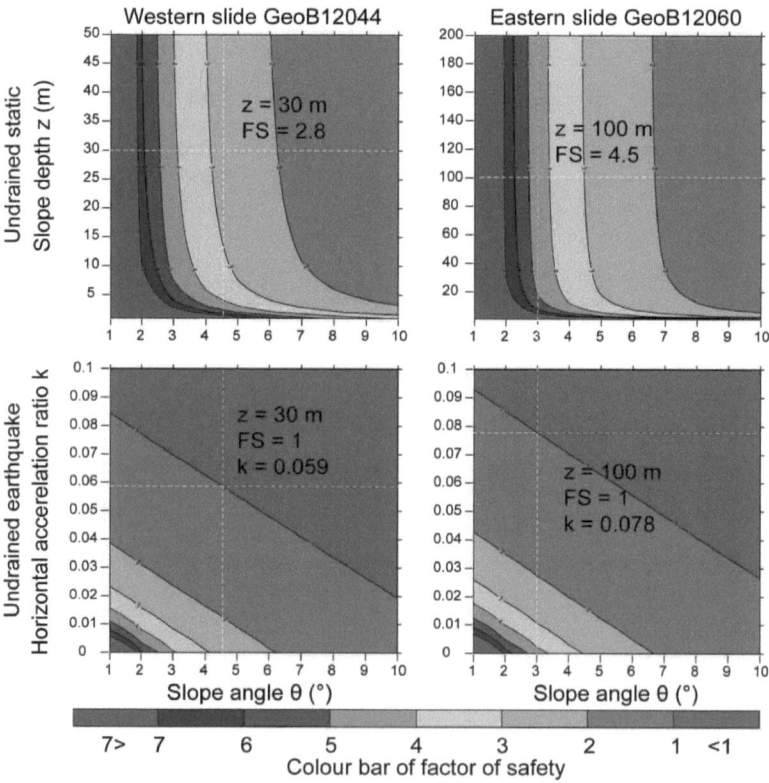

Fig. 4.6 *Undrained slope stability analyses and back-calculations of pseudostatic horizontal acceleration ratio for WS (GeoB12044) and ES (GeoB12060). Dashed white lines indicate current mean values of the parameters for static analysis and values of pseudostatic horizontal acceleration required to trigger slope failure (FS = 1).*

Tab. 4.1 *Parameters used for slope stability calculations (US-Undrained Static, DS-Drained Static, UE-Undrained Earthquake, DE-Drained Earthquake).*

Parameter	WS GeoB12044				ES GeoB12060			
	US	DS	UE	DE	US	DS	UE	DE
S_u (kPa)	1.46z+3.3		32.5		1.7z+5.0		260.0	
z (m)	1-50		30		1-300		100	
θ(°)	1-10				1-10			
c' (kPa)	-	5	-	5	-	9	-	9
φ' (°)	-	32	-	32	-	30	-	30
γ (kN/m²)	-		17.1		-	-	17.4	
λ*	-	0.38	-	0.38	-	0	-	0
k	-		0-0.1	0-0.3	-	-	0-0.1	0-0.3
γ' (kN/m²)	7.31				7.62			
g (m/s²)	9.81				9.81			
FS/k	2.8/-	>7/-	1/0.06	1/0.14	4.5/-	>7/-	1/0.08	1/0.23

4.5 Discussion

4.5.1 Preconditioning factors of WS and ES: superficial failure vs. deep-seated failure

Bathymetric data, seismic profiles and consolidation test results indicate WS is affected by superficial failures with shallow headwall (~30 m vertical displacement) while ES shows deep-seated failure with deeper scars (~100 m). Previous studies in the Ligurian margin have shown that the slope angle is a governing factor for sediment failure (e.g., Cochonat et al. 1993; Migeon et al. 2011). High slope angles promote regular small-volume failure events, which prevent the area to build a thick, potentially unstable sediment package. Sedimentation rates are assumed to be higher near WS on the flank of Var Upper Valley than in the ES region due to regular sediment supply by hyperpycnal flows (Klaucke et al. 2000). On the other hand, hyperpycnal flows are also involved in the gradual undercutting at the base of canyon walls leading to local oversteepening (Migeon et al. 2011). High slope angles with high sedimentation rates and effect of hyperpycnal flows promote superficial failure in WS whereas relative lower slope angle, lower sedimentation rates and without reworking by bottom currents promote the accumulation of a thick but more stable sediment succession in the ES region. The latter then serves as a prerequisite and sufficient material resource for deep-seated failure and larger volumes of slid material.

4.5.2 The influence of earthquake to the slope stability

Superficial failures frequently occur in oversteepened, underconsolidated sediments resulting from high sedimentation rates, while deep-seated failures probably require external constraints such as seismic loading on the sediments to induce slope instability. When considering acceleration-induced earthquakes as a static parameter, it is reasonable to assume a drained pseudostatic model (Mulder et al.,

1994). Critical pseudostatic acceleration as the average equivalent uniform shear stress imposed by seismic shaking represents ~65% of the effective seismic peak ground acceleration (PGA) (Strasser et al. 2011). WS is more vulnerable in undrained conditions where a PGA of 0.09 g (PGA = 0.06 g/0.65) is sufficient to fail the slope. In our study, PGA has been estimated using an empirical attenuation equation after Bindi et al. (2011) (Fig. 4.7). Over the past 30 years, earthquakes with magnitudes 3.8-4.6 have occurred around the Marcel Fault in distances as small as 10 km to WS and 5.6 km to ES (Larroque et al. 2012 and Fig. 4.1B). Despite this short epicentral distance, PGA induced by the M4.6 2001 earthquake (0.03 g for WS and 0.05 g for ES) is still insufficiently strong to trigger instability in either WS or ES. The attenuation relationship indicates that moderate earthquake activity of M5.0 on the spot or stronger earthquakes (e.g., M = 6.1) in epicentral distances < 15 km are required to fail the WS slope. In the ES area, moderate earthquakes with M 5.3 on the spot or > M6.5 earthquakes at distances < 15 km are required to trigger slope failure. From Mulder et al. 1994, PGA ranging from 0.095 g to 0.26 g could be expected for earthquakes with return periods ranging from 100 to 1000 years, respectively. We propose that seismic triggers may have been required for the deep-seated failure in the ES area, but certainly also affected the instability of superficial failure in the WS region.

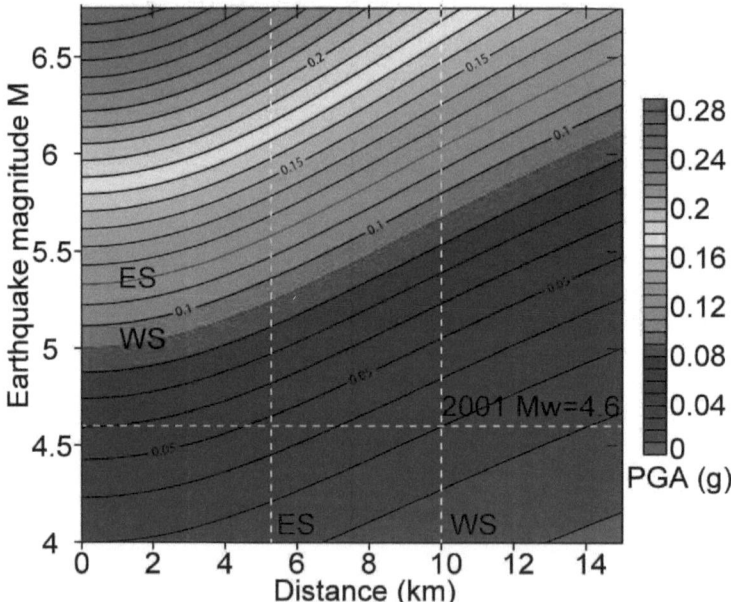

Fig. 4.7 PGA estimates using an empirical attenuation equation after Bindi et al. (2009). Dashed white lines indicate the 2001 earthquake and the distances between the slope scarps of WS and ES to the epicenter of the 2001 earthquake. Red lines indicate the PGA needed to currently trigger slope failure at WS and ES.

4.6 Conclusions

In summary, we have demonstrated how geotechnical properties of sediments and slope stability analysis of two distinct types of slope failure (superficial failure and deep-seated failure) control the Ligurian Margin. Consolidation test results can be used to calculate the amount of sediment removed by slope failure, which is consistent with depth estimates from seismic profiles. The slope angle seems to have a stronger influence on slope instability than slope failure depth below seafloor. For better assessment the potential instability in this tectonic active area, dating of different failure events is mandatory to correlate these data to real seismic events. However, the risk assessment shows that a large-size failure only requires moderate earthquake magnitudes, similar to (or even lower than) those of the 1887 historical event. Given the societal loss associated with a tsunamigenic landslide at the French Riviera, more detailed work has to be carried out in this direction.

Acknowledgments

We thank the captain and crew of the RV Meteor for their support during the cruise M 73/1. This study is funded through DFG-Research Center/Cluster of Excellence "The Ocean in the Earth System" as well as the Chinese Scholarship Council. We also like to acknowledge the anonymous reviewers for their constructive remarks.

References

Bindi D, Pacor F, Luzi L, Puglia R, Massa M, Ameri G, Paolucci R (2011) Ground motion prediction equations derived from the Italian strong motion database. Bulletin of Earthquake Engineering. 9 (6): 1899-1920.

Cochonat P, Bourillet JF, Savoye B, Dodd L (1993) Geotechnical characteristics and instability of submarine slope sediments, the nice slope (N-W Mediterranean Sea). Marine Georesources & Geotechnology. 11 (2): 131-151.

Dan G, Sultan N, Savoye B (2007) The 1979 Nice harbour catastrophe revisited: Trigger mechanism inferred from geotechnical measurements and numerical modelling. Marine Geology. 245 (1-4): 40-64.

Förster A, Spieß V, Kopf AJ, Dennielou B (2010) Mass Wasting Dynamics at the Deeper Slope of the Ligurian Margin (Southern France). Submarine Mass Movements and Their Consequences Advances in Natural and Technological Hazard Research. Springer, Dordrecht, Heidelberg, London, New York. pp. 66-77.

Locat J, Lee HJ (2002) Submarine landslides: advances and challenges. Canadian Geotechnical Journal. 39 (1): 193-212.

Klaucke I, Cochonat P (1999) Analysis of past seafloor failures on the continental slope off Nice (SE France). Geo-Marine Letters. 19 (4): 245-253.

Klaucke I, Savoye B, Cochonat P (2000) Patterns and processes of sediment dispersal on the continental slope off Nice, SE France. Marine Geology. 162 (2-4): 405-422.

Kopf A, Cruise Participants (2008) Report and Preliminary Results of Meteor Cruise M 73/1: LIMA-LAMO (Ligurian Margin Landslide Measurements & Observatory). Berichte Fachbereich Geowissenschaften, Universität Bremen, 264. 161p.

Larroque C, Scotti O, Ioualalen M (2012) Reappraisal of the 1887 Ligurian earthquake (western Mediterranean) from macroseismicity, active tectonics and tsunami modelling. Geophysical Journal International. 190 (1): 87-104.

Leynaud D, Sultan N (2010) 3-D slope stability analysis: A probability approach applied to the nice slope (SE France). Marine Geology. 269 (3-4): 89-106.

Migeon S, Cattaneo A, Hassoun V, Larroque C, Corradi N, Fanucci F, Dano A, Mercier de Lepinay B, Sage F, Gorini C (2011) Morphology, distribution and origin of recent submarine landslides of the Ligurian Margin (North-western Mediterranean): some insights into geohazard assessment. Marine Geophysical Research. 32 (1-2): 225-243.

Morgenstern N (1967) Submarine slumping and the initiation of turbidity currents. Marine geotechnique. 189-220.

Mulder T, Tisot J-P, Cochonat P, Bourillet J-F (1994) Regional assessment of mass failure events in the Baie des Anges, Mediterranean Sea. Marine Geology. 122 (1-2): 29-45.

Silva AJ, LaRosa P, Brausse M, Baxter CD, Bryant WR (2001) Stress states of marine sediments in plateau and basin slope areas of the northwestern Gulf of Mexico. Offshore Technology Conference.

Stegmann S, Sultan N, Kopf A, Apprioual R, Pelleau P (2011) Hydrogeology and its effect on slope stability along the coastal aquifer of Nice, France. Marine Geology. 280 (1-4): 168-181.

Strasser M, Hilbe M, Anselmetti F (2011) Mapping basin-wide subaquatic slope failure susceptibility as a tool to assess regional seismic and tsunami hazards. Marine Geophysical Research. 32 (1-2): 331-347.

Sultan N, Cochonat P, Canals M, Cattaneo A, Dennielou B, Haflidason H, Laberg JS, Long D, Mienert J, Trincardi F, Urgeles R, Vorren TO, Wilson C (2004) Triggering mechanisms of slope instability processes and sediment failures on continental margins: a geotechnical approach. Marine Geology. 213 (1-4): 291-321.

5 Conclusion and outlook

5.1 Conclusion

This thesis aimed to improve our understanding of the initiation of submarine landslides in different geological settings. Different slope failures events from passive continental margin (Uruguayan and northern Argentine margin) and active continental margins (Gela Basin and Ligurian margin) were chosen to analyze preconditioning factors and triggering mechanisms of submarine landslides. The thesis demonstrates the integration of geological, geophysical, sedimentological, physical and geotechnical information with combination of infinite slope stability and pseudostatic earthquake analysis studying the submarine landslide. On the basis of the results from this study the following major conclusions can be drawn:

Uruguayan and northern Argentine margin: passive continental margin

The Uruguayan and northern Argentine margin is characterized by a pattern of steep scarps on the gentle slope (slope angle: ~1.7-2.2°) offshore Uruguay and slope failure at the steep canyon headwall and flanks (slope angle: ~ 22°) with mass transport deposits stacked at the canyon mouth. The slope offshore Uruguay is stable under the present sedimentary condition which dominated by clayey silt material with joint roles of fluvial discharge of the Rio de la Plata River and low-energy contour current. Slope failures are expected to occur for moderate earthquakes (M 4) in the NS are or strong events (e.g., M = 7) in epicentral distances < 15 km. In contrast, the slope offshore northern Argentina has a low stability under present sedimentary condition which characterized by silty sand material with downslope driven processes and extensive reworking of strong contour currents. Small-scale slope failures occur both during static conditions and certainly during infrequent seismic events such as moderate earthquakes in the near-field or strong earthquake (e.g., M = 7) in the far-field (epicentral distance < 45 km). The comparison of slope failure modes between open slope area (the slope offshore Uruguay) and adjacent canyon area (the slope offshore northern Argentina) reveals that different oceanographic and sedimentary settings result in different styles submarine mass movements, while earthquake as additional triggering factor are important to be considered even in passive margin settings.

Gela Basin, the central Mediterranean continental margin: active margin

The Gela Basin is characterized by widespread occurrence of repeated submarine mass movements. Two small-scales slope failure events (Northern Twin Slide and Southern Twin slide) are exposed by fresh headwalls and a set of failed sediments dipping at 1.5-4.5°. Strongly underconsolidated of the sediments in the upslope of Northern Twin Slide mainly attribute to rapid sedimentation in fine sediments. Despite of the high overpressure presented in the sediments, the stability analysis suggest the slope is stable under present conditions. Slope failure may be triggered by moderate seismic shaking (M4.0-M4.8) in the study area, or even strong events if farther away. Geotechnical investigation reveals that overpressure generated by rapid sedimentation rates is important to precondition the slope towards low stability, thus combining with external triggering mechanisms (e.g., earthquake), it could finally initiate slope failure.

Ligurian margin, Southern France: active margin

The deeper slope of Ligurian margin is characterized two distinct types of slope failure: superficial slope failures on the steep flank (slope angle: ~4.6°) of Upper Var Canyon and deep-seated failures on the less steep slope (slope angle: ~3°) eastward Upper Var Canyon. High sedimentation rates and effect of hyperpycnal flows promote superficial failure on the flank of Upper Var Canyon, while lower sedimentation rates and without reworking by bottom currents promote the accumulation of a thick but more stable sediment succession on the slope eastward Upper Var Canyon. Slope stability and earthquake analysis reveals that seismic triggers, i.e. M5.3 earthquakes on the spot, may have been required for the deep-seated failure, but certainly also affected the instability of superficial failure on the flank of Upper Var Canyon.

5.2 Outlook

Although this thesis contributed to a better answering the question: "why do some areas fail whereas adjacent areas do not?" using multi-methodological approach on different types of slope failure in adjacent areas both on the passive margin and active continental margins, some questions remain unsolved and more new questions arose. Some key questions, which have been only partially being answered, are briefly outlined following:

How overpressure changes in the dynamic loading (e.g., Seismic shaking)?

Overpressure is recognized as a key parameter for the assessment of slope stability. In this study, overpressure generated by rapid sedimentation rates is only considered, while other processes to induce overpressure (e.g., dynamic loading, gas charging sediments) are not taken into account. Further studies of in-situ pore pressure measurement, ideally in long-term observations (Stegmann et al. 2011, 2012) and pore pressure change during the seismic loading are needed in order to further investigate the dynamic response and better assess the slope stability.

What is the role of weak layers in slope stability?

Mechanically weak layers are well known to decrease the shear strength of sediments and play important roles in landslide initiation, but this knowledge is often restricted to the shallow sub-seafloor where such layer can be sampled with common coring techniques (e.g., gravity and piston cores). However, it is hard to unambiguously determine where the weak layers are geophysically or to recover them from large depth below seafloor (generally slip surfaces located in scale of ~100 mbsf; McAdoo et al 2000; Urlaub et al. 2012). It is necessary to use advanced geophysical and coring tools to reach the weak layers for better understanding slope failure processes.

When did submarine slope failures occur and what is the frequency of recurrence?

For better understanding the influence of earthquake to submarine landslide, age dating of different (ideally multiple) events and estimation of the recurrence frequency of slope failures are vital to correlate these data to e.g., seismic events, so that a better prediction of future submarine mass movements is possible to build resilient societies.

References

McAdoo BG, Pratson LF, Orange DL (2000) Submarine landslide geomorphology, US continental slope. Marine Geology 169 (1-2): 103-136.

Stegmann S, Sultan N, Garziglia S, Pelleau P, Apprioual R, Kopf A, Zabel M A Long-term Monitoring Array for Landslide Precursors: A Case Study at the Ligurian Slope (Western Mediterranean Sea). In: Offshore Technology Conference, 2012.

Stegmann S, Sultan N, Kopf A, Apprioual R, Pelleau P (2011) Hydrogeology and its effect on slope stability along the coastal aquifer of Nice, France. Marine Geology 280 (1-4): 168-181.

Urlaub M, Zervos A, Talling PJ, Masson DG, Clayton CI (2012) How do~ 2° slopes fail in areas of slow sedimentation? s sensitivity study on the influence of accumulation rate and permeability on submarine slope stability. In:Yamada Y et al. (eds) Submarine Mass Movements and Their Consequences. Andvances in Natural and Technological Hazard Research: 277-287.

Acknowledgements

This four-year Ph.D. project can be carried out with the fund through DFG-Research Center/Cluster of Excellence "The Ocean in the Earth System" as well as the financial support from Chinese Scholarship Council. I would like thanks to all those who help and support throughout the process of completing my Ph.D. thesis.

First, I would like to express my highest gratitude to my supervisor Prof. Dr. Achim Kopf for offering me the opportunity to study in Marine Geotechnics group, MARUM, University of Bremen. With persistent support of Achim and influence of his personality charm, I learn how to be a good researcher and a good man. I am also extremely grateful to my co-supervisor Prof. Dr. Michael Strasser for inspiring me so many ideas and giving me the most immediate support on my thesis. I want to thank my supervisor Prof. Dr. Xiang Wei in China University of Geosciences (Wuhan) for encouraging me in pursuit of knowledge abroad.

I also would like to thank Prof. Dr. Tobias Mörz who agreed to evaluate this thesis. Many thanks also go to all Professors and colleagues of my Ph.D. thesis defense committee.

I would like to thank my co-authors, Dr. Benedict Preu, Prof. Dr. Sebastian Krastel, Dr. Till Hanebuth, Jannis Kuhlmann, Prof. Dr. Katrin Huhn, Dr. Annika Förster, Dr. Sylvia Stegmann, Prof. Dr. Michael Strasser, Prof. Dr. Achim Kopf for inspiring discussions and constructive revisions that great improvement the quality of the manuscripts in this thesis.

I am extremely grateful to all my colleagues in the groups of Marine Geotechnics and Marine Engineering Geology. A special thank goes to Alois Steiner for the scientific discussions. I am obliged to Matthias Lange, Daniel Otto, Andre Hüppers, Matt Ikari and Guvain Wiemer for the technical support and quick help for laboratory tests. Many thanks for Ina Schulz, Anna Reusch, Maria Belke Brea and Johannes Hüsener for help on geotechnical tests. I would like to thank Petra Renken, Chritian Zollner and Marc Huhndorf, Franziska Hellmich for their help making my life as an international student easier and enjoyable.

I would like to thank all the friends for the fantastic time we had in Bremen.

And last but not least I would like express my deep gratitude to my parents, my brother, my wife and my son for their never ending love, support and encouragement.

Appendix A: Core descriptions, physical and geotechnical properties of Uruguayan and northern Argentine margin

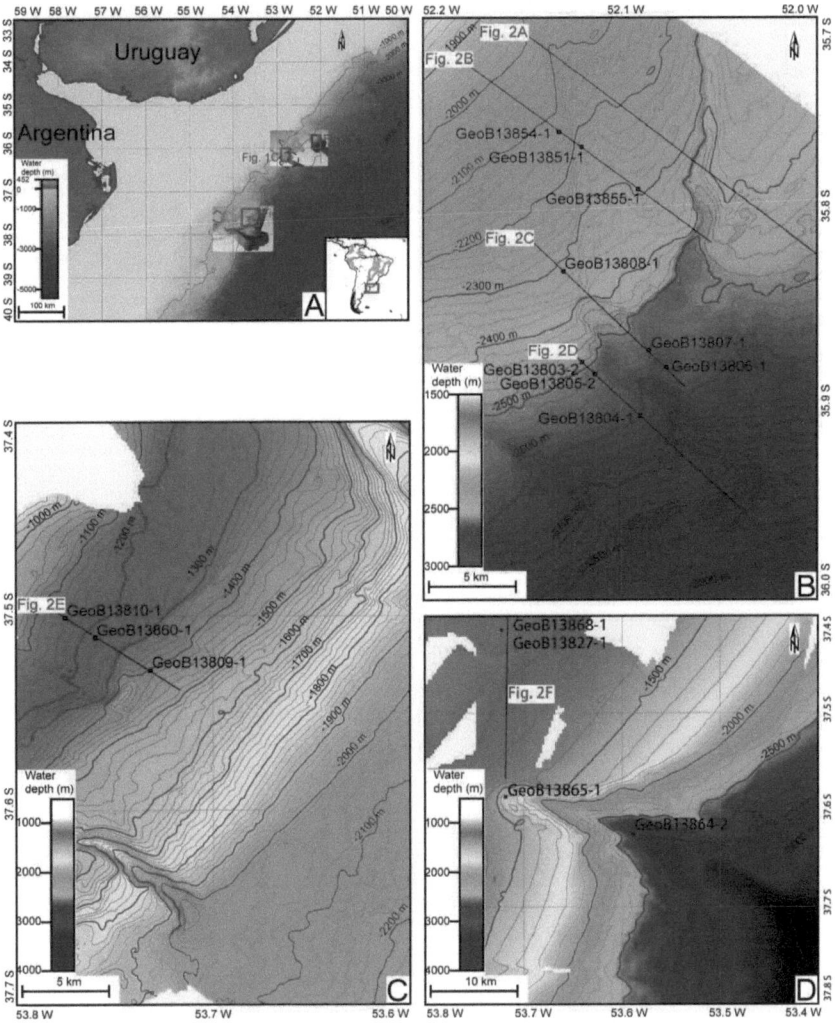

Fig. A1 *(A) Map showing the location of the study area along the Uruguayan and northern Argentine margin. (B) and (C) Bathymetric contour maps of Northern slide(NS). Black lines indicate the position of Parasound and seismic profiles. Black dots indicate core locations. (D) Bathymetric map of Southern Canyon (SC). Black line indicates the position of seismic profiles. Black dots indicate core locations.*

Appendix A: Core descriptions, physical and geotechnical properties of Uruguayan and northern Argentine margin

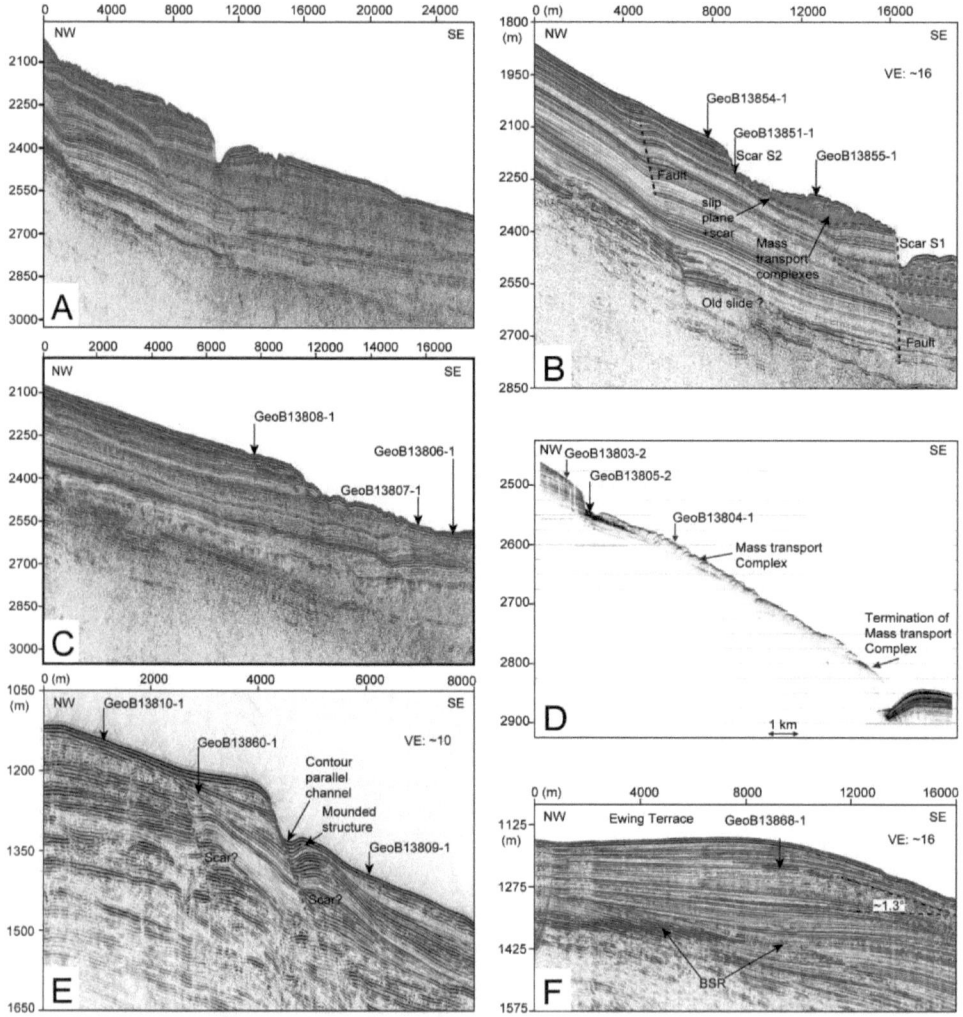

Fig. A2 *(A), (B), (C) and (E) Seismic profiles of NS. (D) Parasound profile of the NS. (F) Seismic profiles of SC.*

Appendix A: Core descriptions, physical and geotechnical properties of Uruguayan and northern Argentine margin

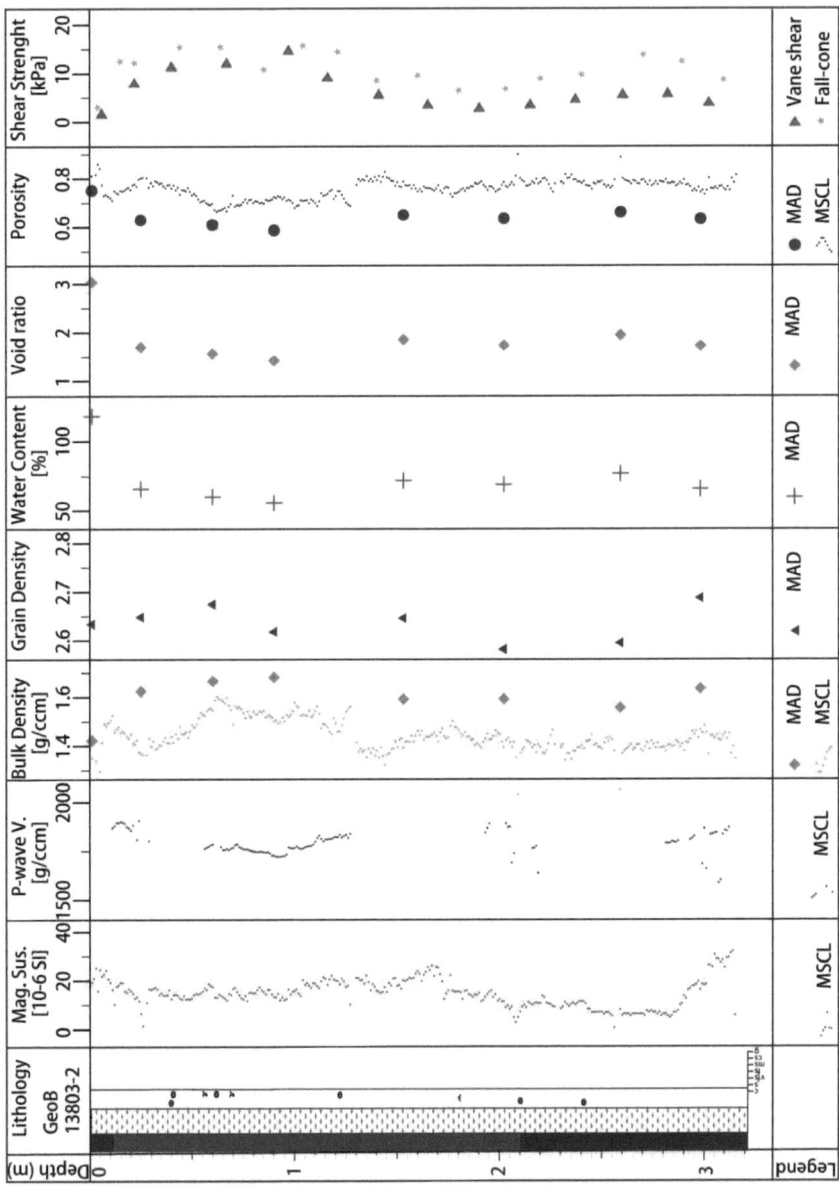

Fig. A3 *Core descriptions, sediment physical and geotechnical properties of GeoB13802-1.*

Appendix A: Core descriptions, physical and geotechnical properties of Uruguayan and northern Argentine margin

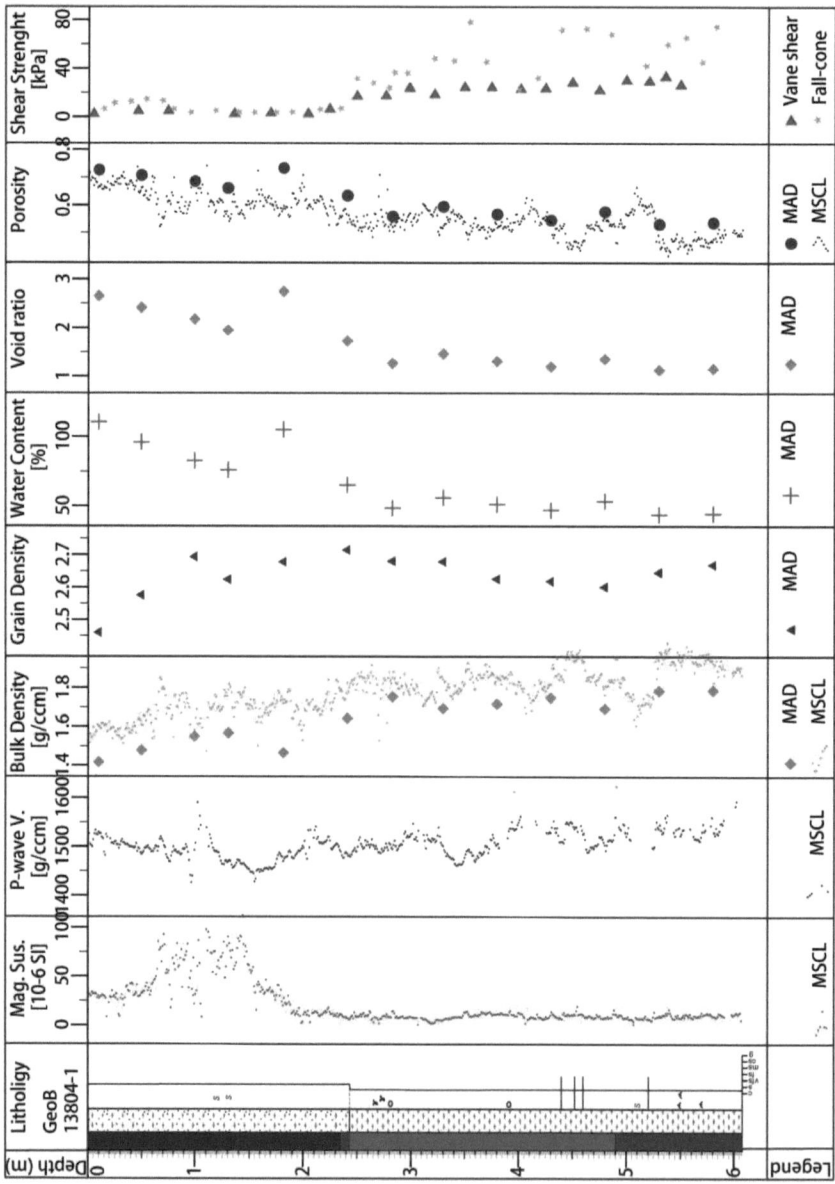

Fig. A4 Core descriptions, sediment physical and geotechnical properties of GeoB13804-1.

Appendix A: Core descriptions, physical and geotechnical properties of Uruguayan and northern Argentine margin

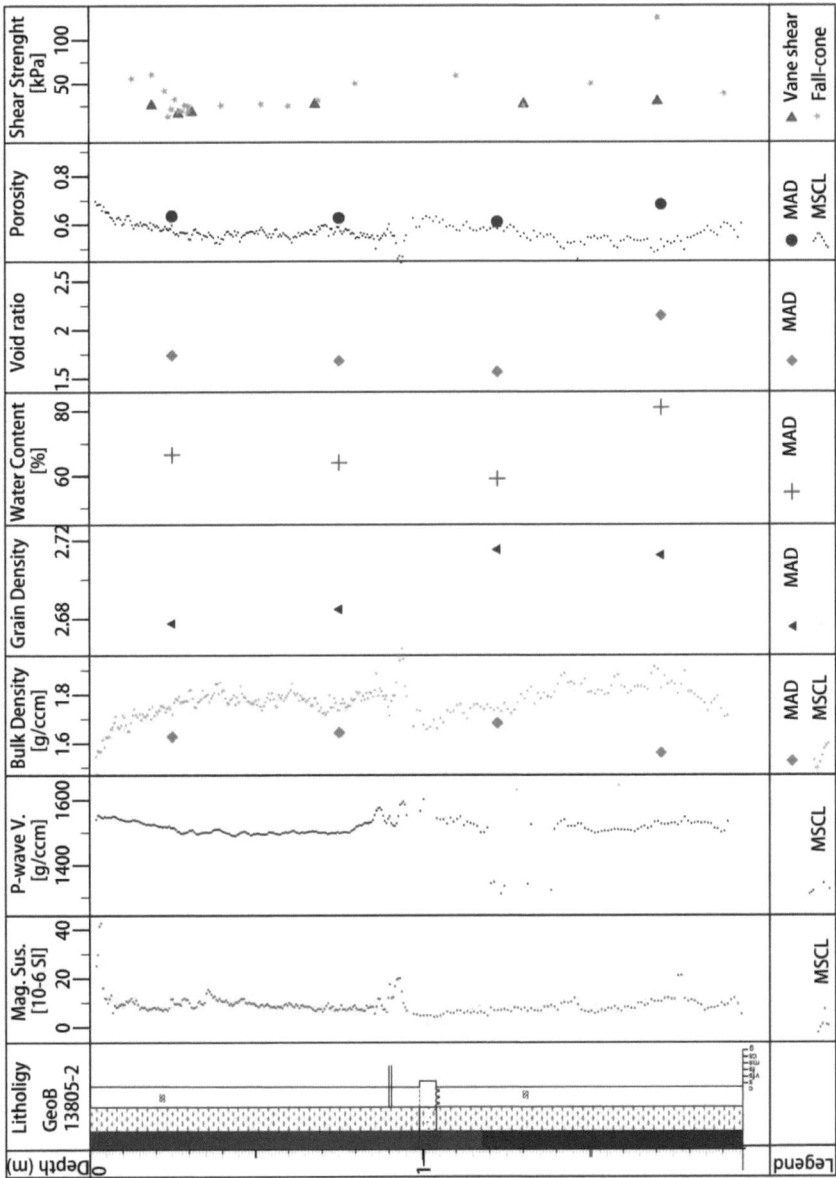

Fig. A5 Core descriptions, sediment physical and geotechnical properties of GeoB13805-2.

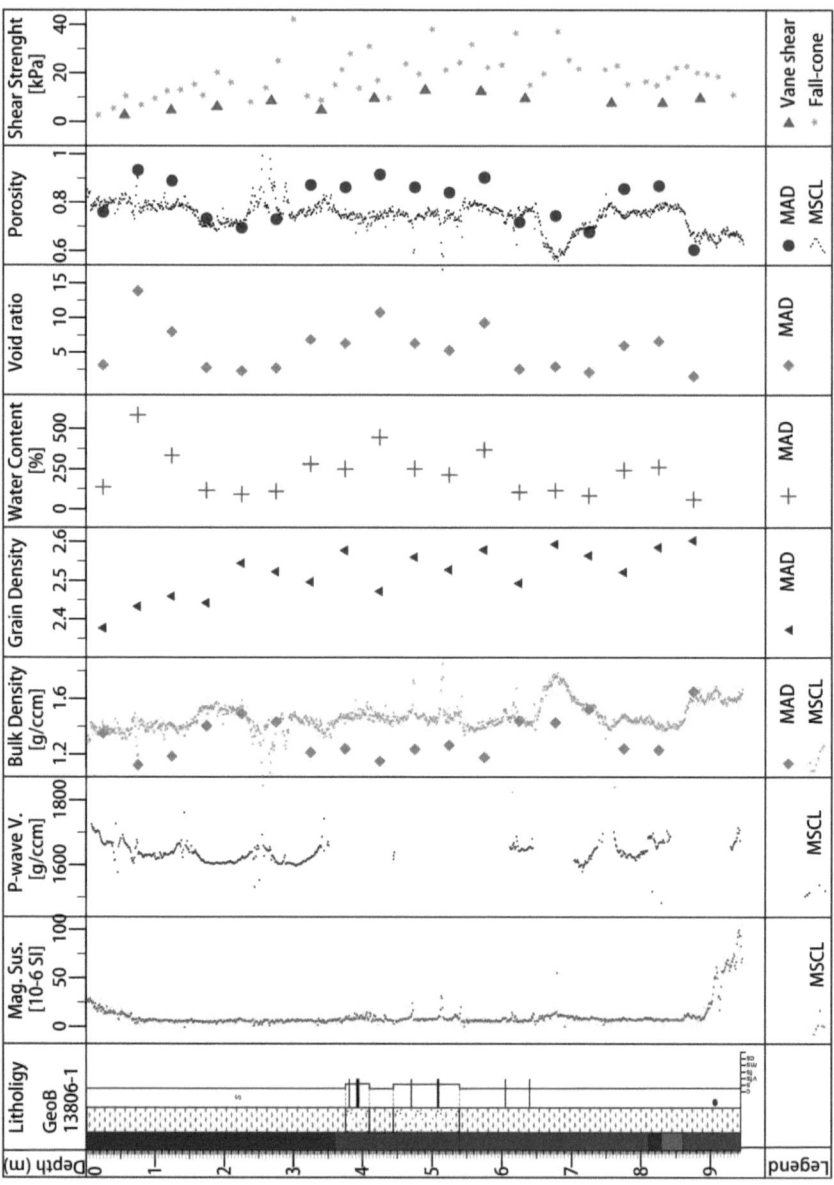

Fig. A6 *Core descriptions, sediment physical and geotechnical properties of GeoB13806-1.*

Appendix A: Core descriptions, physical and geotechnical properties of Uruguayan and northern Argentine margin

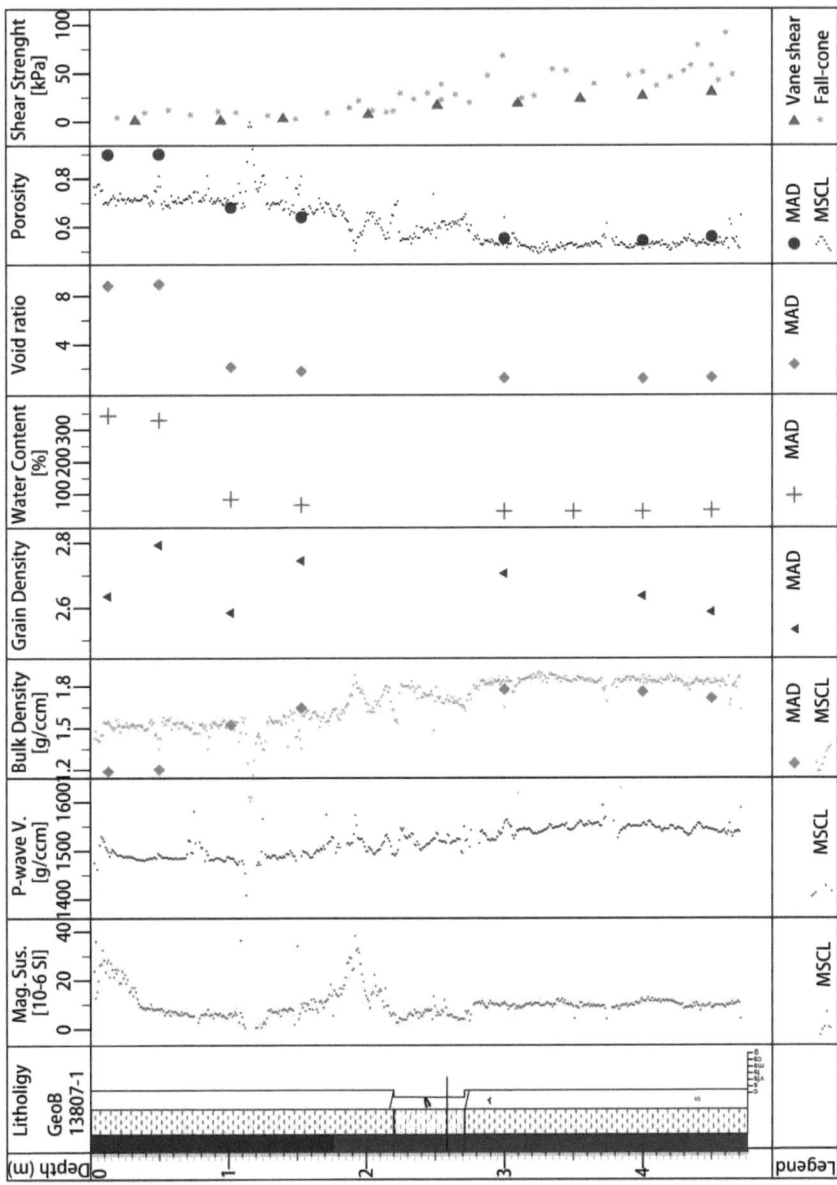

Fig. A7 *Core descriptions, sediment physical and geotechnical properties of GeoB13807-1.*

Appendix A: Core descriptions, physical and geotechnical properties of Uruguayan and northern Argentine margin

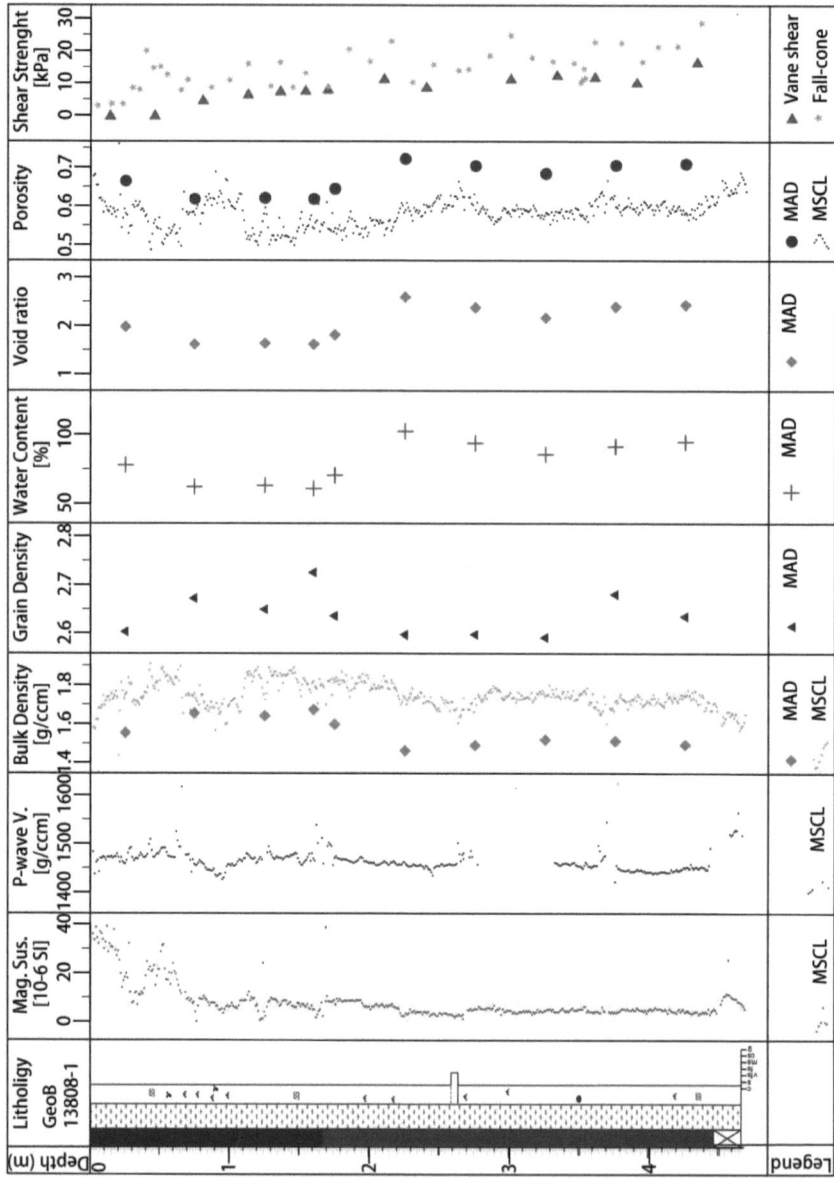

Fig. A8 *Core descriptions, sediment physical and geotechnical properties of GeoB13808-1.*

Appendix A: Core descriptions, physical and geotechnical properties of Uruguayan and northern Argentine margin

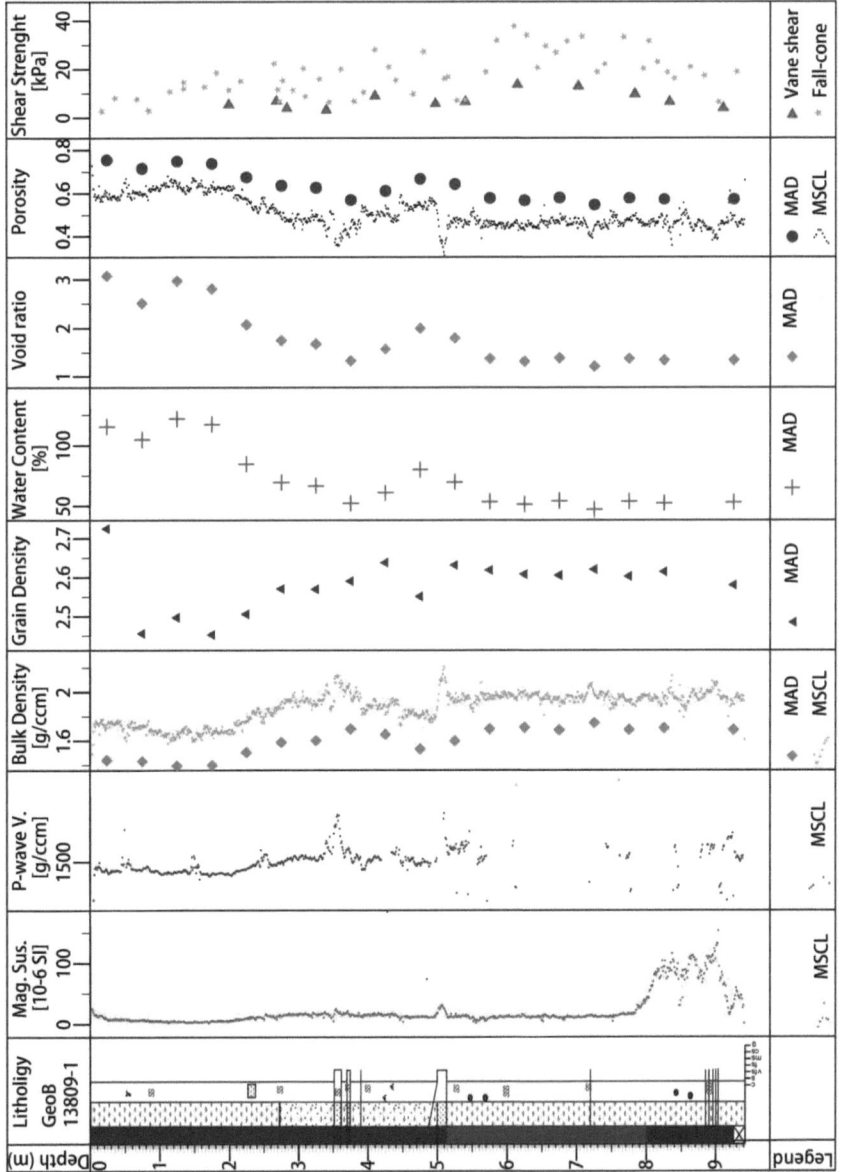

Fig. A9 Core descriptions, sediment physical and geotechnical properties of GeoB13809-1.

Appendix A: Core descriptions, physical and geotechnical properties of Uruguayan and northern Argentine margin

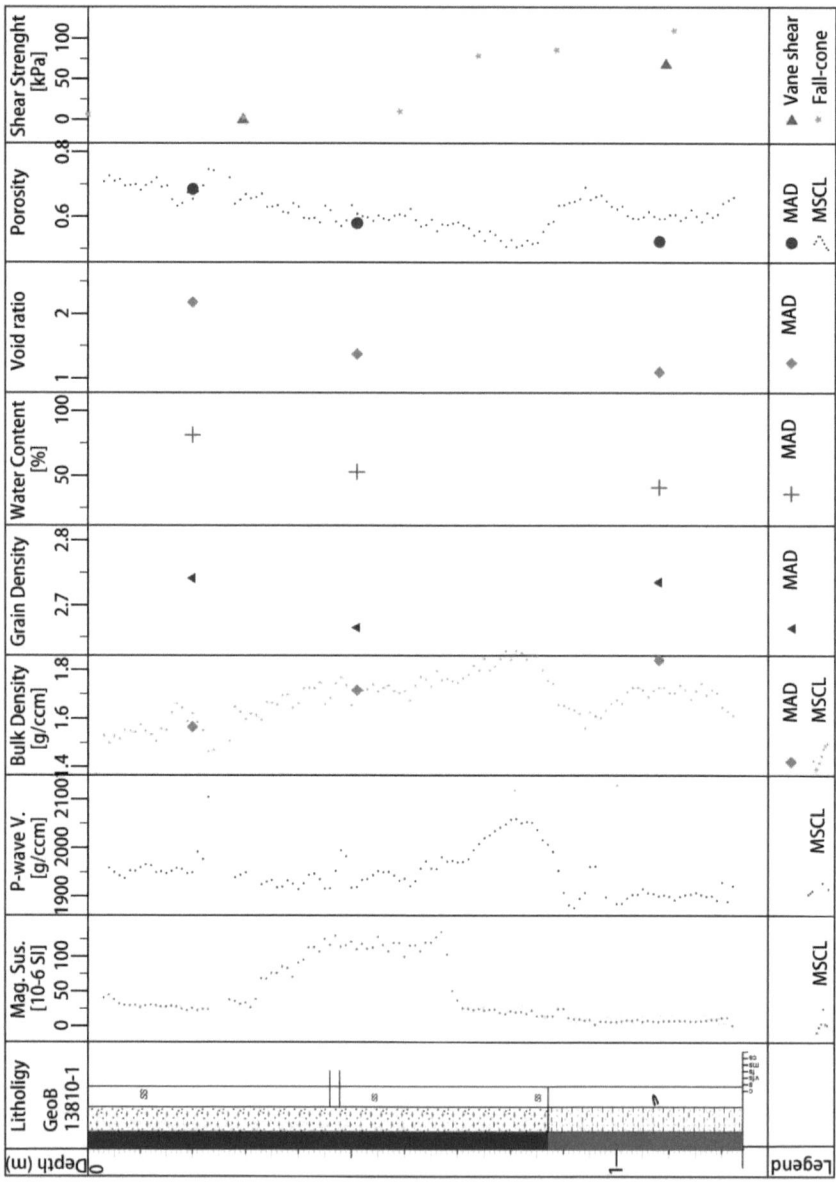

Fig. A10 *core descriptions, sediment physical and geotechnical properties of GeoB13810-1.*

Appendix A: Core descriptions, physical and geotechnical properties of Uruguayan and northern Argentine margin

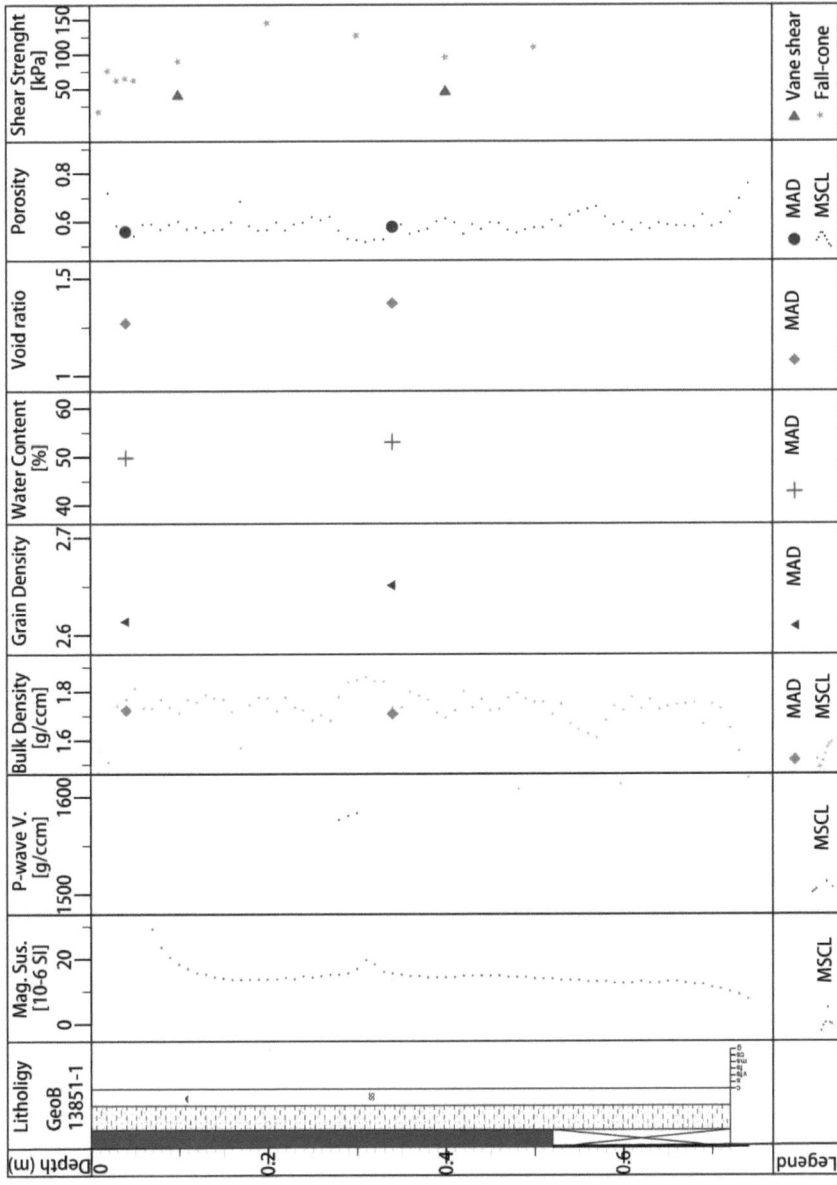

Fig. A11 *Core descriptions, sediment physical and geotechnical properties of GeoB13851-1.*

Appendix A: Core descriptions, physical and geotechnical properties of Uruguayan and northern Argentine margin

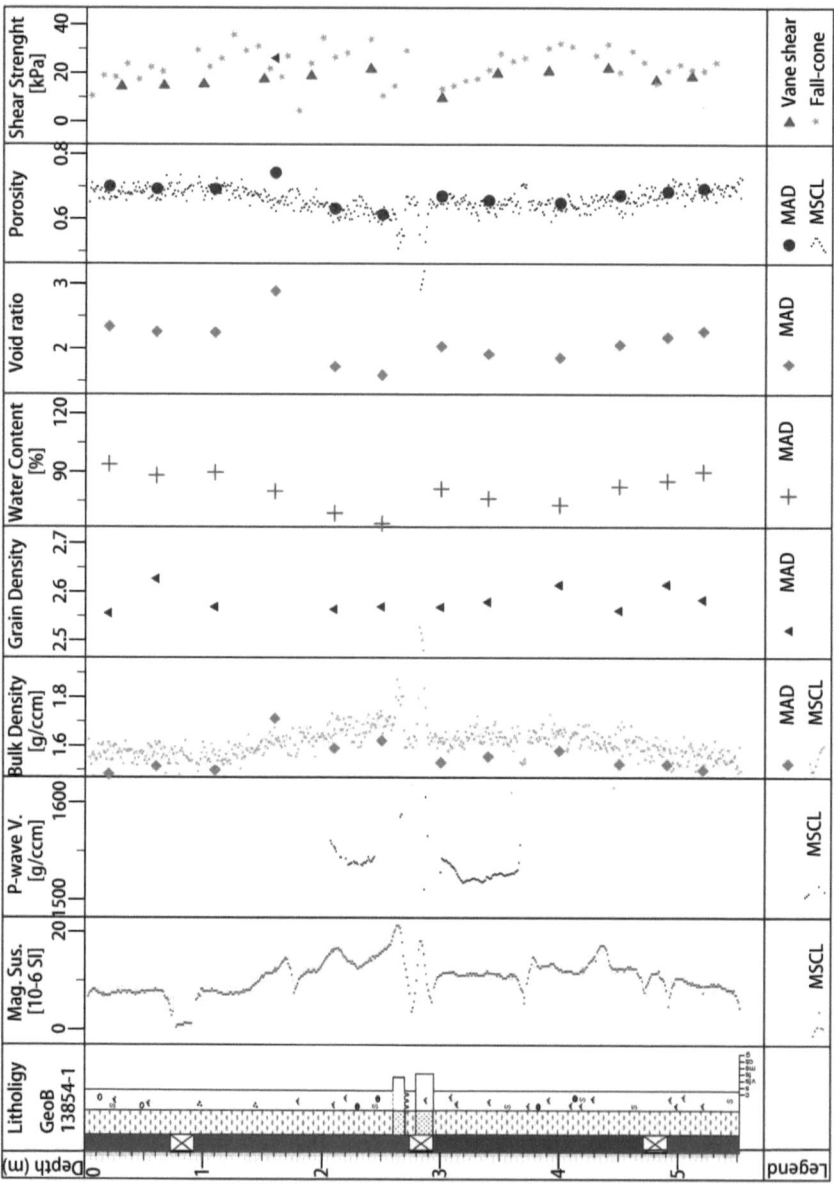

Fig. A12 *Core descriptions, sediment physical and geotechnical properties of GeoB13854-1.*

Appendix A: Core descriptions, physical and geotechnical properties of Uruguayan and northern Argentine margin

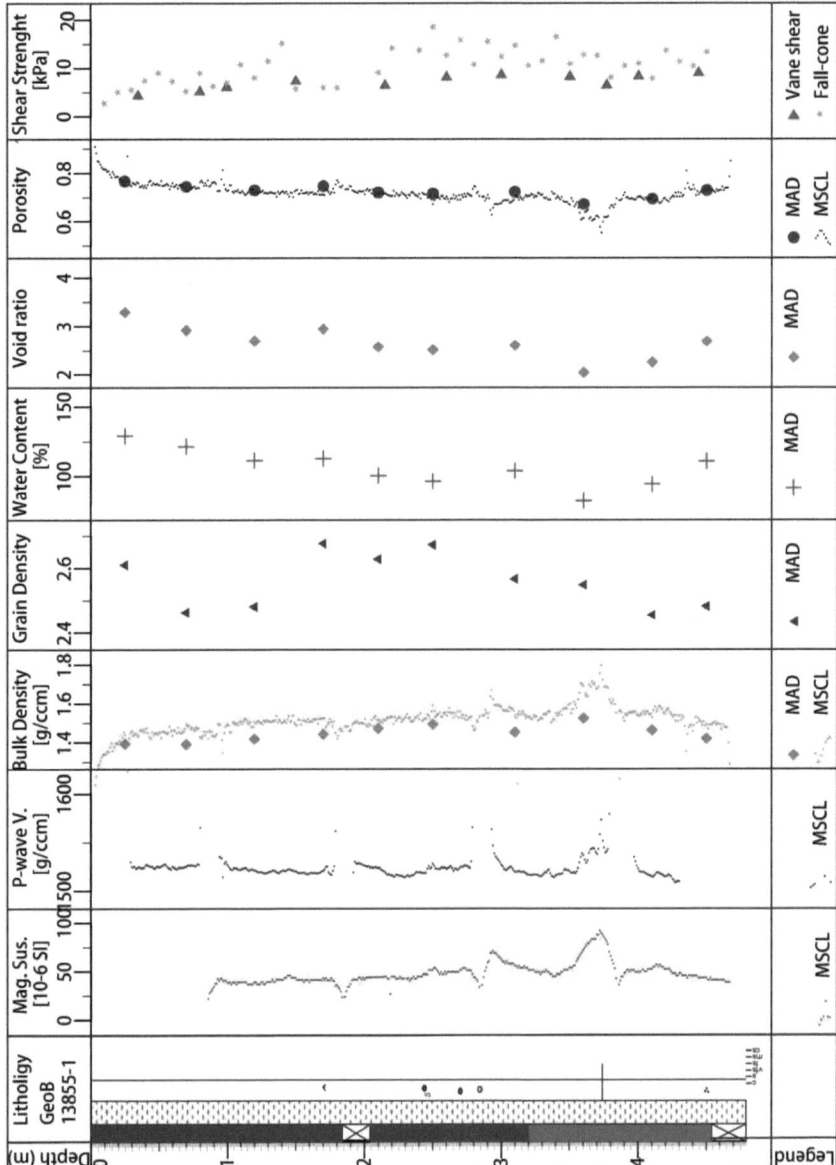

Fig. A13 Core descriptions, sediment physical and geotechnical properties of GeoB13855-1.

Appendix A: Core descriptions, physical and geotechnical properties of Uruguayan and northern Argentine margin

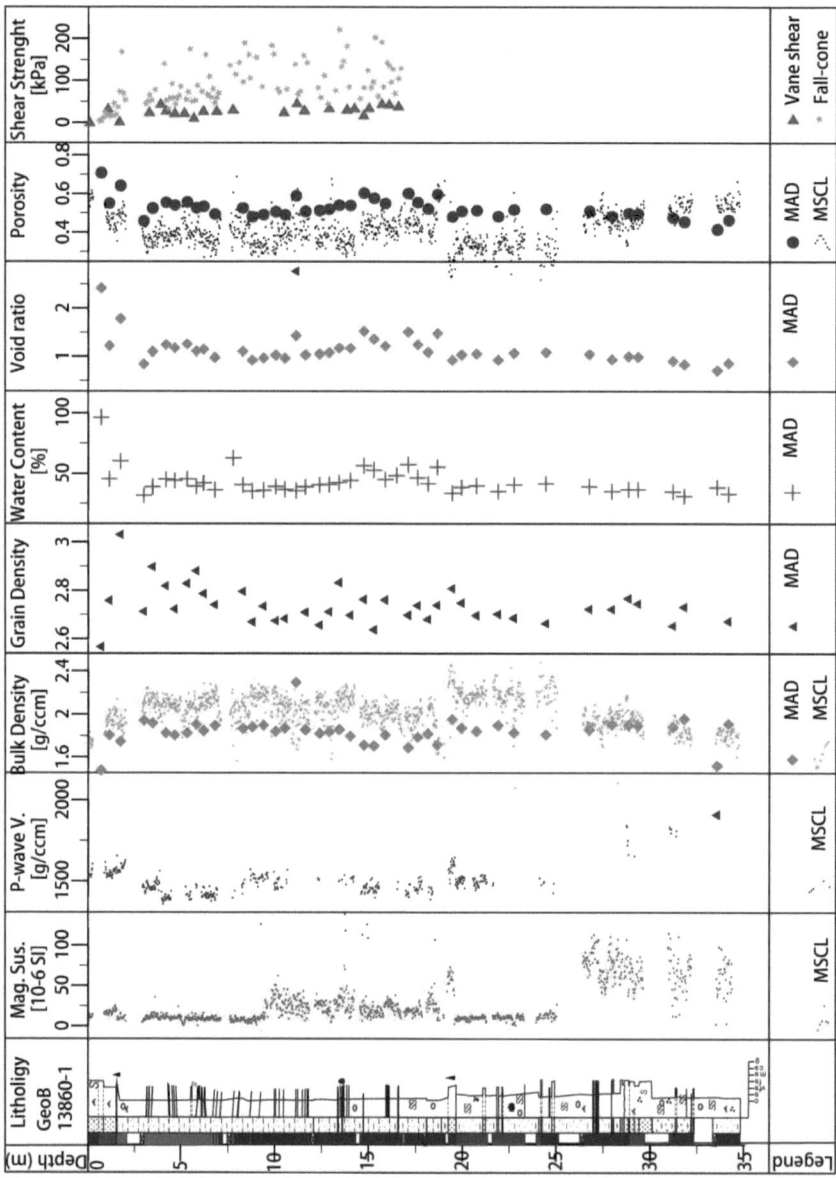

Fig. A14 *Core descriptions, sediment physical and geotechnical properties of GeoB13860-1.*

Appendix A: Core descriptions, physical and geotechnical properties of Uruguayan and northern Argentine margin

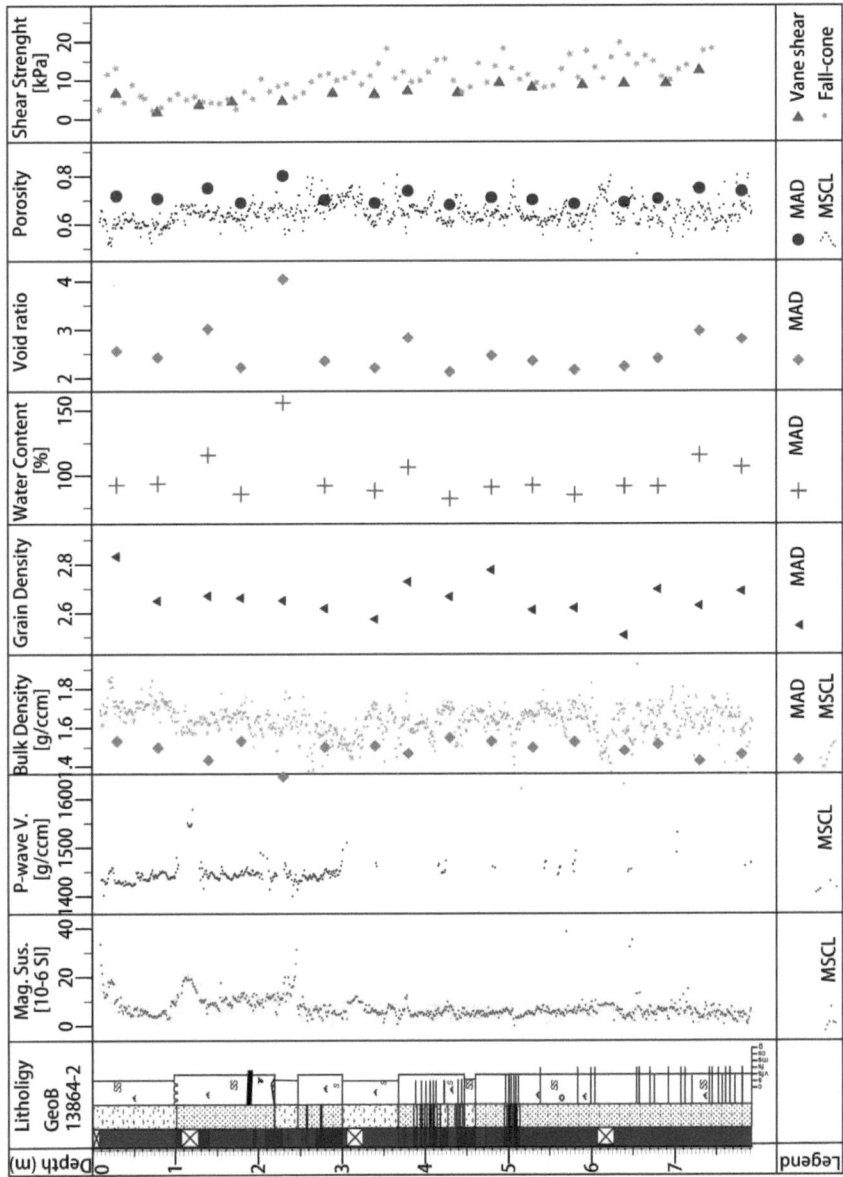

Fig. A15 Core descriptions, sediment physical and geotechnical properties of GeoB13864-1.

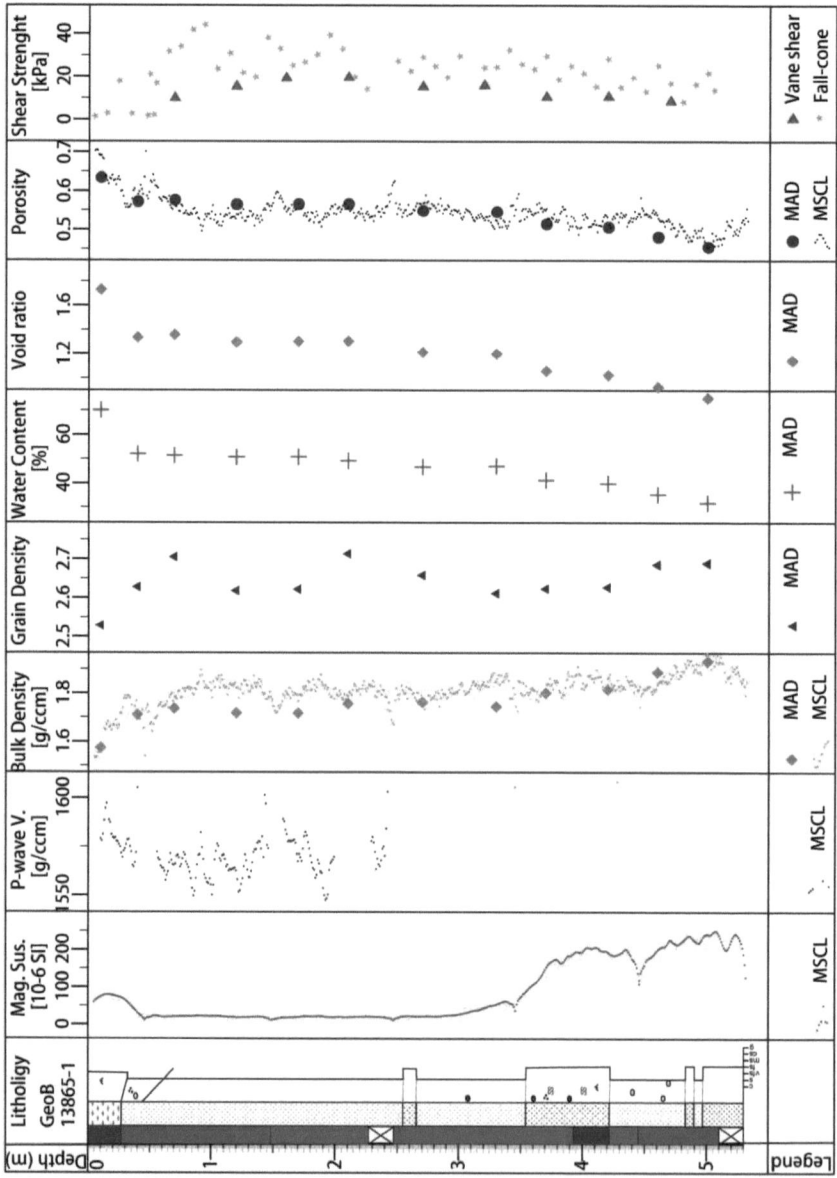

Fig. A16 Core descriptions, sediment physical and geotechnical properties of GeoB13865-1.

Appendix A: Core descriptions, physical and geotechnical properties of Uruguayan and northern Argentine margin

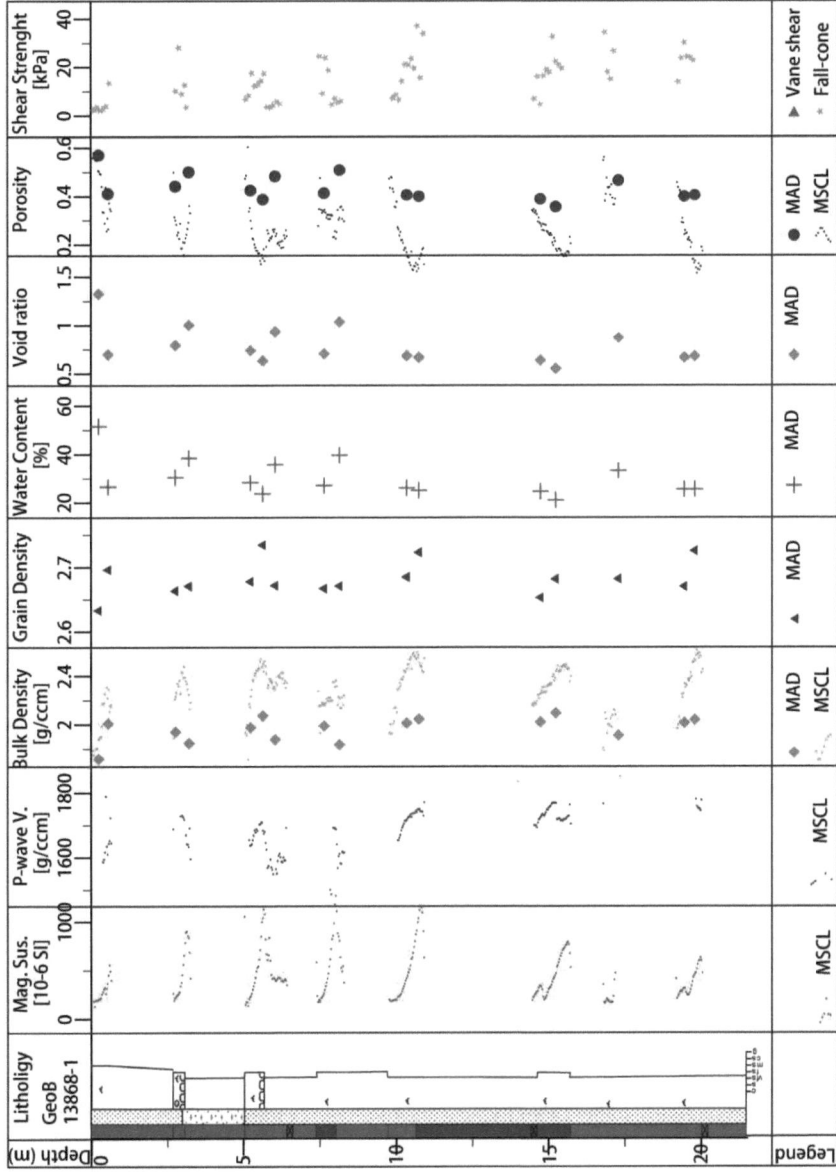

Fig. A17 Core descriptions, sediment physical and geotechnical properties of GeoB13868-1.

Appendix B: Core descriptions, physical and geotechnical properties of Gela Basin

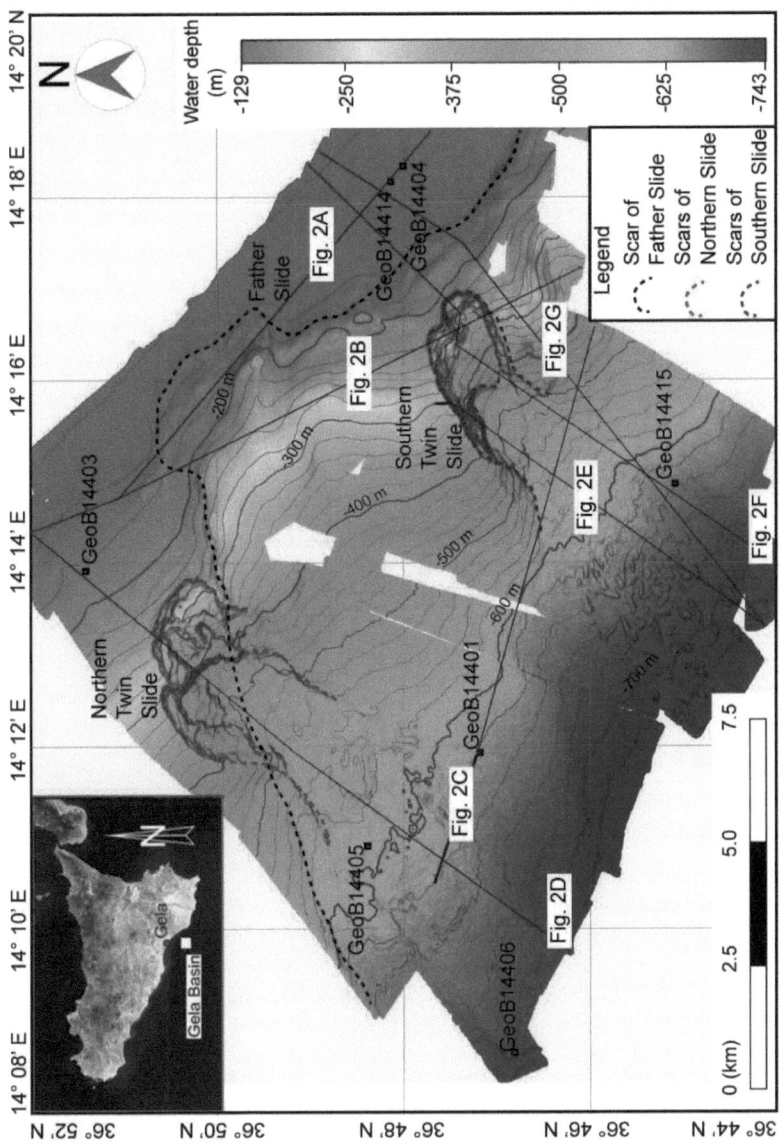

Fig. B1 *Bathymetric map of the Twins slides in the Gela Basin offshore Sicily. Black dots indicate core locations. Black lines indicate the locations of Parasound profiles presented in Fig. B2.*

Appendix B: Core descriptions, physical and geotechnical properties of Gela Basin

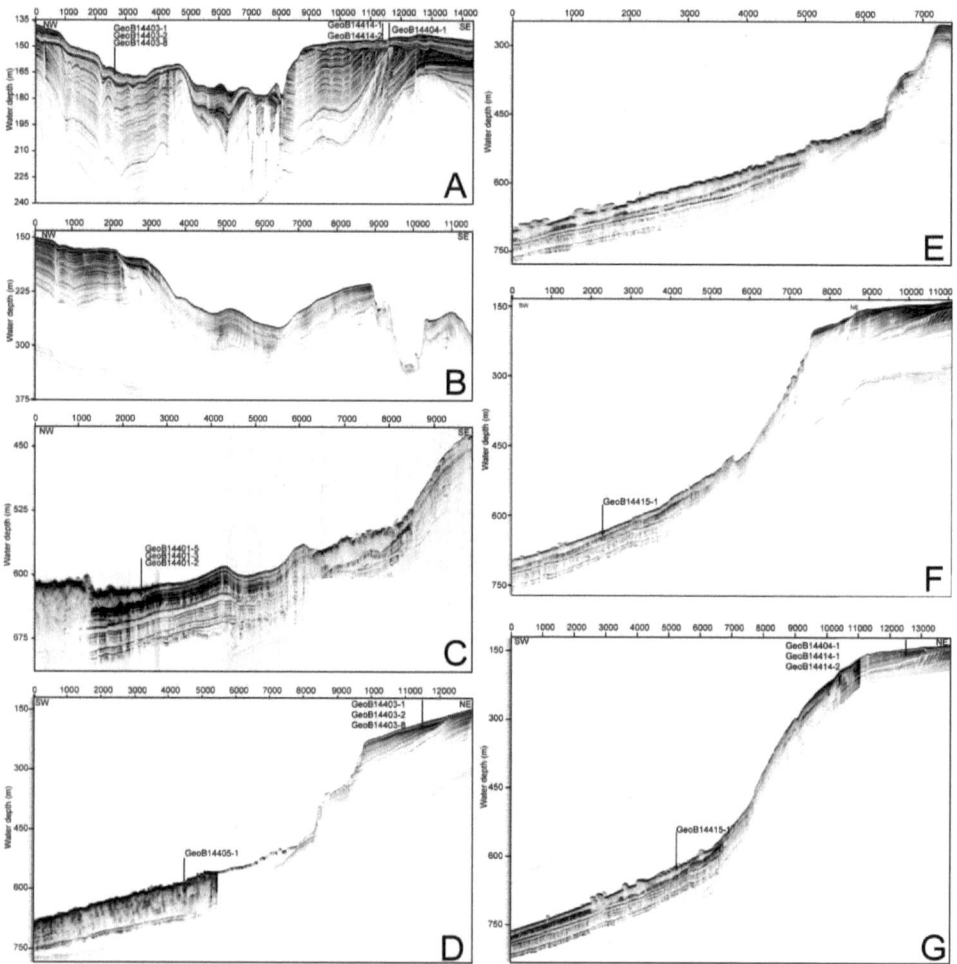

Fig. B2 Parasound profiles of the Twins Slides. (A) Parasound profile crosses the upslopes of NTS (Northern Twin Slide) and STS (Southern Twin Slide). (B) Parasound profile crosses the upslope of NTS and the headwall of STS. (C) Parasound profile crosses the deposit areas of NTS and STS. (D) Parasound profile crosses along NTS. (E), (F), (G) Parasound profile crosses along STS.

Appendix B: Core descriptions, physical and geotechnical properties of Gela Basin

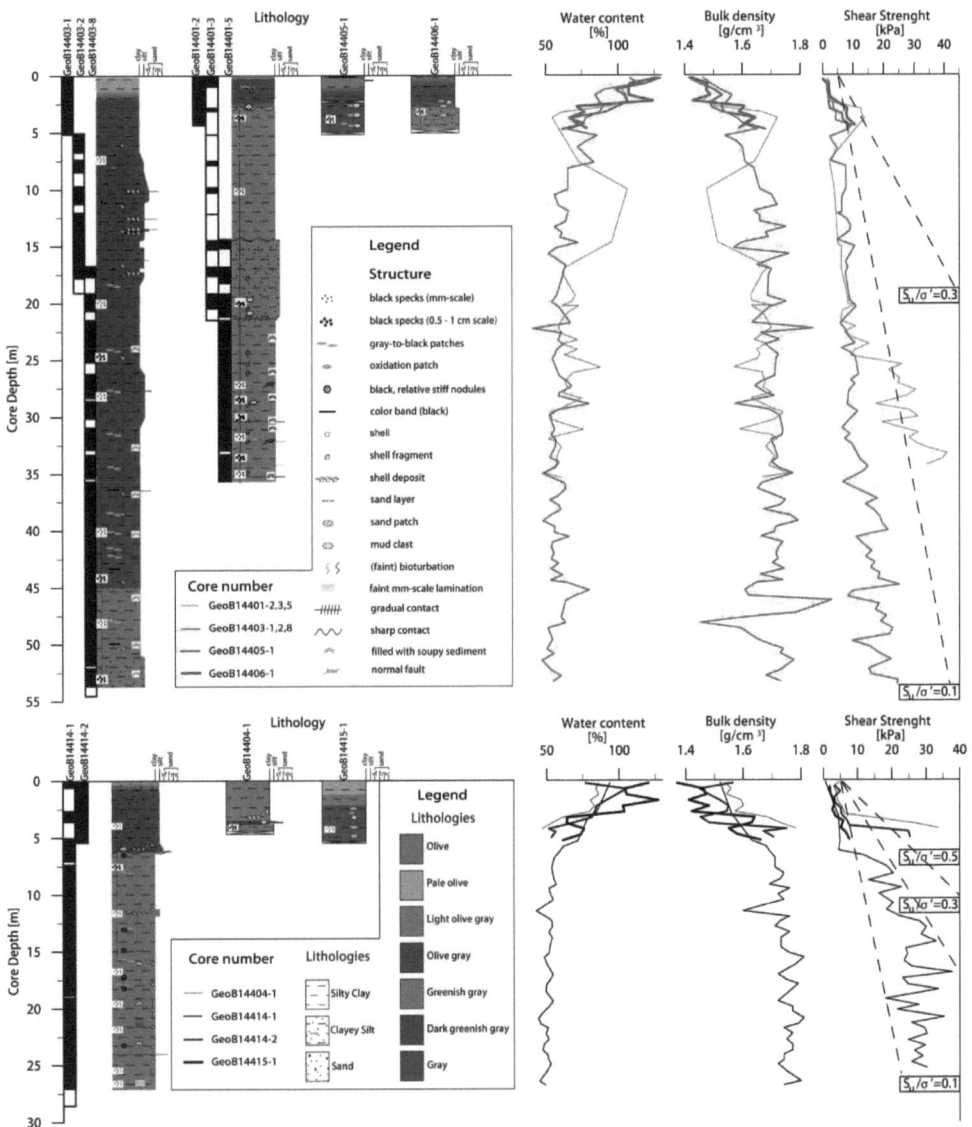

Fig. B3 Core descriptions, physical and geotechnical properties of cores in Gela Basin.

Appendix B: Core descriptions, physical and geotechnical properties of Gela Basin

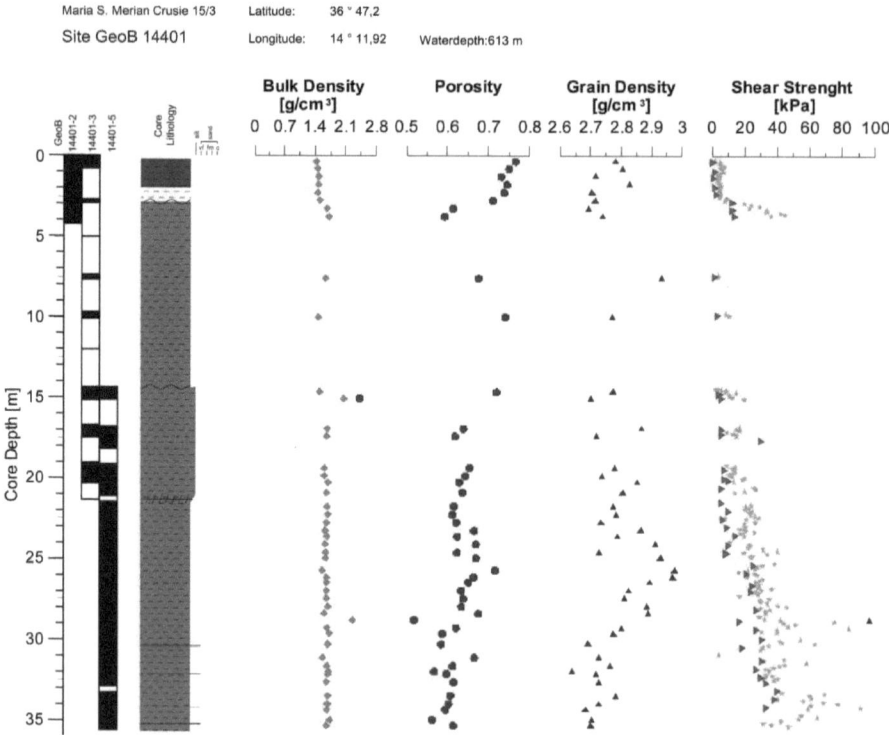

Fig. B4 Core descriptions, physical and geotechnical properties of GeoB14401.

Appendix B: Core descriptions, physical and geotechnical properties of Gela Basin

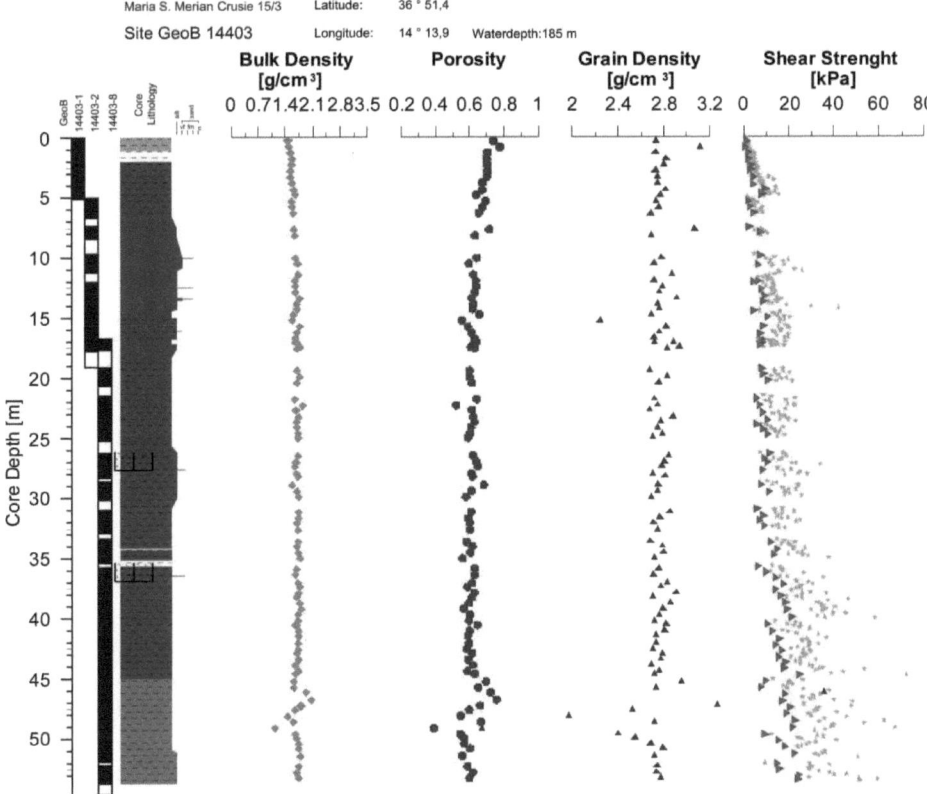

Fig. B5 Core descriptions, physical and geotechnical properties of GeoB14403.

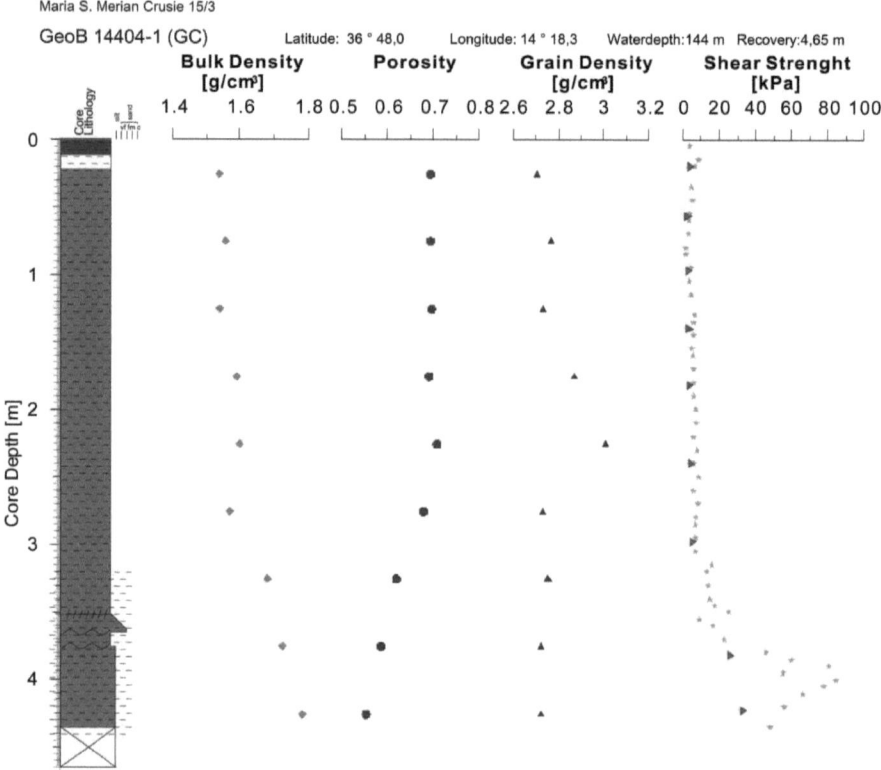

Fig. B6 *Core descriptions, physical and geotechnical properties of GeoB14404-1.*

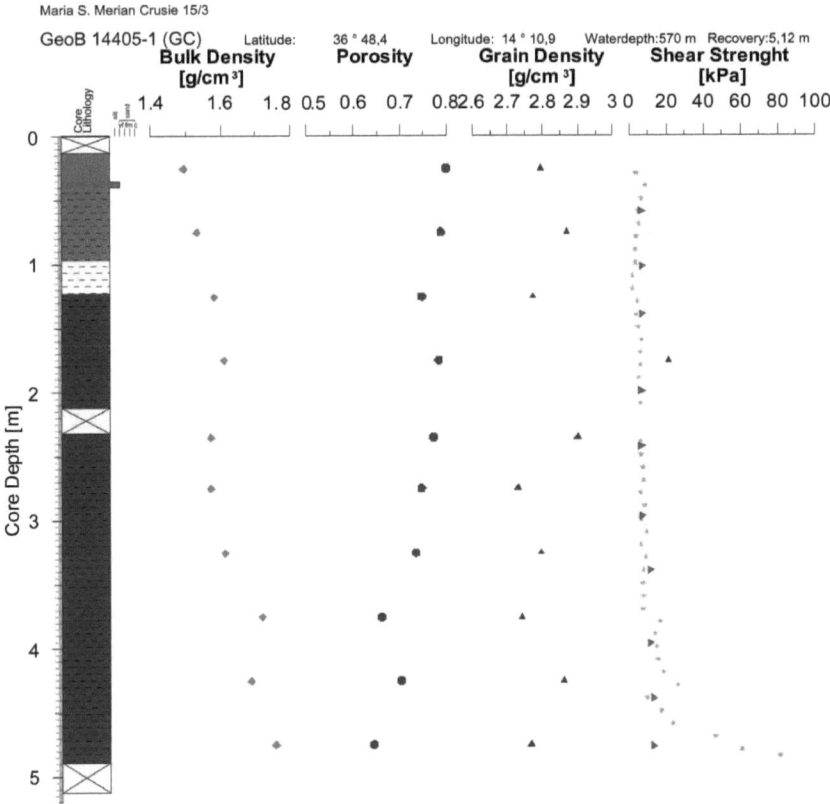

Fig. B7 Core descriptions, physical and geotechnical properties of GeoB14405-1.

Appendix B: Core descriptions, physical and geotechnical properties of Gela Basin

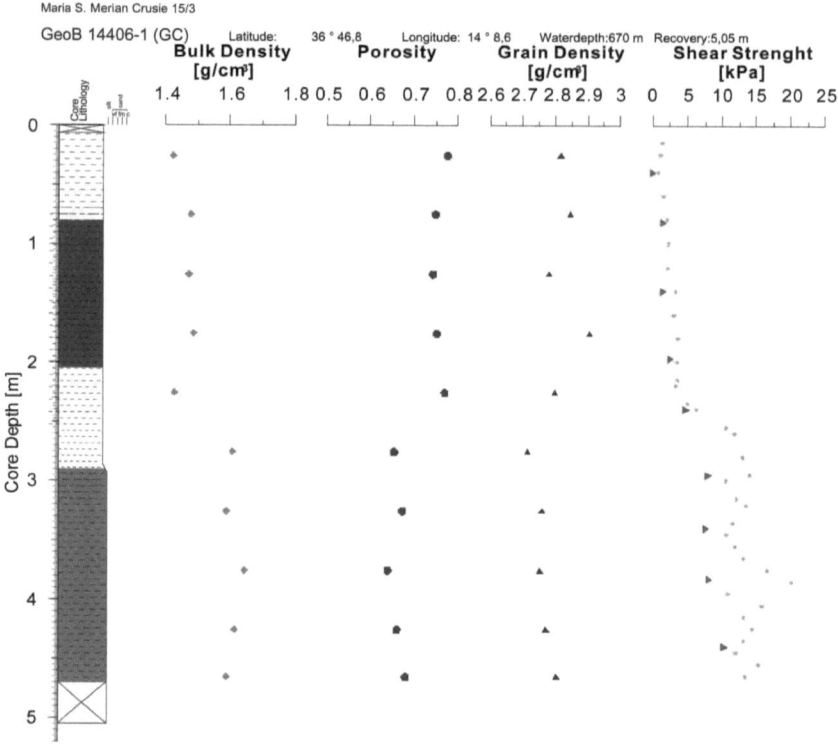

Fig. B8 *Core descriptions, physical and geotechnical properties of GeoB14406-1.*

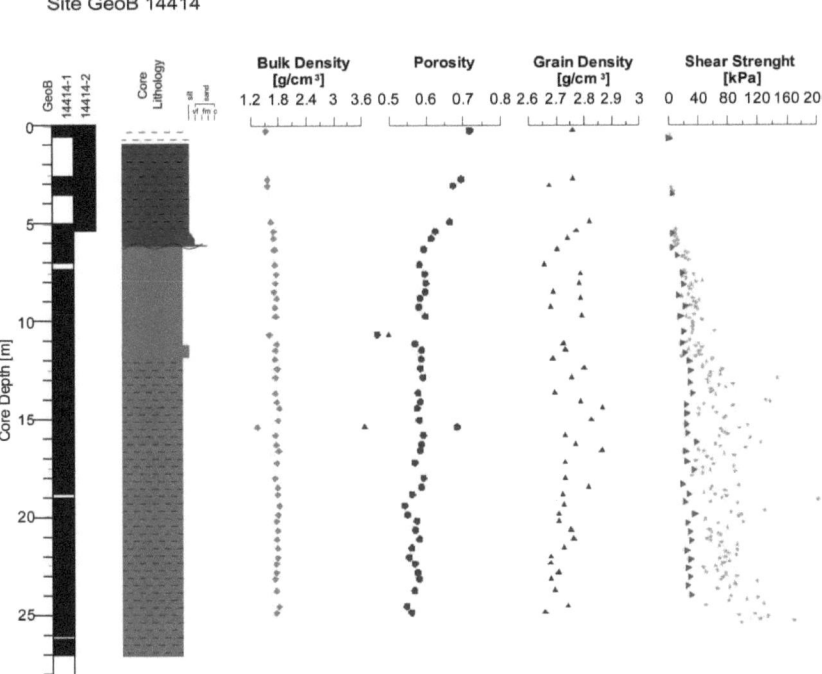

Fig. B9 Core descriptions, physical and geotechnical properties of GeoB14414.

Appendix B: Core descriptions, physical and geotechnical properties of Gela Basin

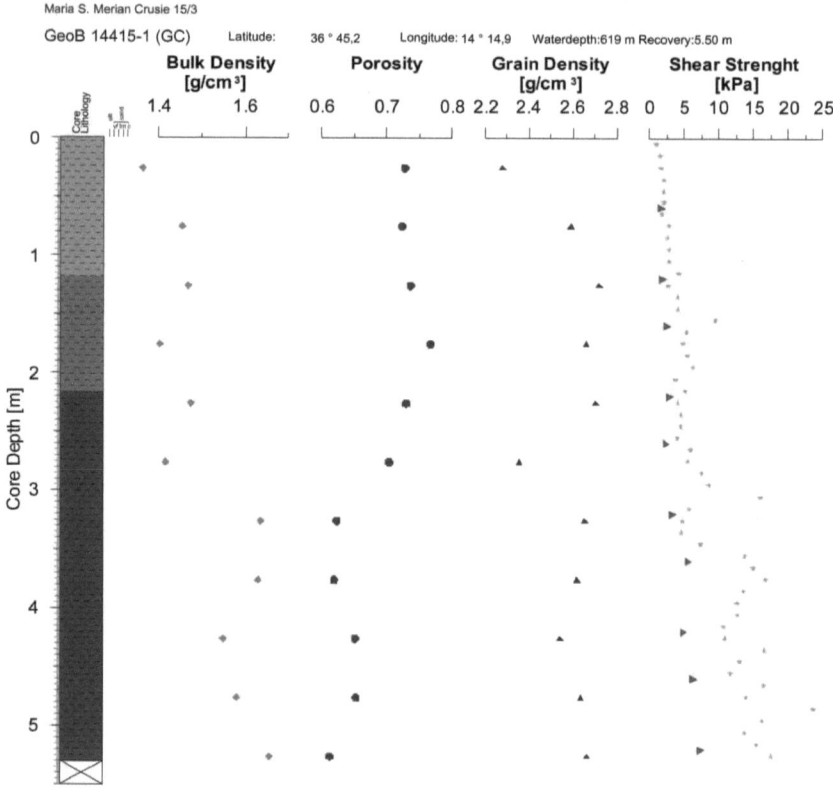

Fig. B10 Core descriptions, physical and geotechnical properties of GeoB14401.

I want morebooks!

Buy your books fast and straightforward online - at one of the world's fastest growing online book stores! Environmentally sound due to Print-on-Demand technologies.

Buy your books online at
www.get-morebooks.com

Kaufen Sie Ihre Bücher schnell und unkompliziert online – auf einer der am schnellsten wachsenden Buchhandelsplattformen weltweit!
Dank Print-On-Demand umwelt- und ressourcenschonend produziert.

Bücher schneller online kaufen
www.morebooks.de

OmniScriptum Marketing DEU GmbH
Heinrich-Böcking-Str. 6-8
D - 66121 Saarbrücken
Telefax: +49 681 93 81 567-9

info@omniscriptum.com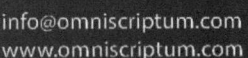
www.omniscriptum.com

Printed by Books on Demand GmbH, Norderstedt / Germany